Studies in Computational Intelligence

Volume 640

Series editor

Janusz Kacprzyk, Polish Academy of Sciences, Warsaw, Poland
e-mail: kacprzyk@ibspan.waw.pl

About this Series

The series "Studies in Computational Intelligence" (SCI) publishes new develop-ments and advances in the various areas of computational intelligence—quickly and with a high quality. The intent is to cover the theory, applications, and design methods of computational intelligence, as embedded in the fields of engineering, computer science, physics and life sciences, as well as the methodologies behind them. The series contains monographs, lecture notes and edited volumes in computational intelligence spanning the areas of neural networks, connectionist systems, genetic algorithms, evolutionary computation, artificial intelligence, cellular automata, self-organizing systems, soft computing, fuzzy systems, and hybrid intelligent systems. Of particular value to both the contributors and the readership are the short publication timeframe and the worldwide distribution, which enable both wide and rapid dissemination of research output.

More information about this series at http://www.springer.com/series/7092

Theodor Borangiu · Damien Trentesaux
André Thomas · Duncan McFarlane
Editors

Service Orientation in Holonic and Multi-Agent Manufacturing

Springer

Editors
Theodor Borangiu
Faculty of Automatic Control and Computer
  Science
University Politehnica of Bucharest
Bucharest
Romania

Damien Trentesaux
University of Valenciennes and
  Hainaout-Cambresis
Valenciennes
France

André Thomas
University of Lorraine
Épinal
France

Duncan McFarlane
Institute for Manufacturing Engineering
  Department
Cambridge University
Cambridge
UK

ISSN 1860-949X   ISSN 1860-9503   (electronic)
Studies in Computational Intelligence
ISBN 978-3-319-30335-2   ISBN 978-3-319-30337-6   (eBook)
DOI 10.1007/978-3-319-30337-6

Library of Congress Control Number: 2016933325

Printed on acid-free paper

This Springer imprint is published by Springer Nature
The registered company is Springer International Publishing AG Switzerland

Foreword

We are living today in an extraordinary period of growing complexity of manu-facturing systems. This is, as a matter of fact, the result of technological develop-ment in several areas:

- Computing systems are becoming more powerful every day. This is a conse-quence of Moore's law which is still valid: the key parameters of HW equip-ment is doubled (processing speed and memory capacity) or reduced to half (energy consumption) every 18 months. The SW architectures and tools are also evolving at a rapid pace and succession.
- Communication systems are becoming more broadband, faster, more efficient, and are built to support the idea of cyber-physical systems.
- Automation methodologies are becoming more intelligent, their development being oriented towards decentralized intelligent systems organized as a com-munity of autonomous units without any central element. AI principles are being massively deployed and exploited.

Especially the field of distributed intelligent systems has influenced manufac-turing and production management quite strongly. Let us summarize briefly the short history of this development to better understand the state of the art and trends in a broader context; this will also underline the importance of scientific events such as SOHOMA.

The first ideas of distributed intelligent systems appeared in connection with the holonic visions 25 years ago. The very first pilot implementations led to the development of the first holons—in principle reactive agents—and to introduction of the first standard in the field (IEC 16499). Once engineers intended to add intelligence to the distributed elements we started to talk about agents, agent-based systems and multi-agent systems (MAS). Specific platforms to run multi-agent systems with specialized functionalities like yellow pages, white pages, brokers, sniffers, etc. evolved being supported by the FIPA Association; then, FIPA com-munication and architecture standards appeared. FIPA efforts are rather frozen at the moment, but the FIPA standards accepted 10–12 years ago are still in use and accepted in the field of MAS.

The MAS philosophy applied to industrial control allowed thinking about and conceiving new approaches and solutions. Progressively, products and semi-products started to be represented by SW agents that were able to communicate, negotiate and coordinate their activities, not only in manufacturing and transport processes. The products became active elements during their execution life cycle. The PROSA-like way of thinking strongly influenced the field, causing that not only semi-products but also humans were considered as resources represented by agents. This might be considered as a significant technology breakthrough in the field of decentralized control and production management.

But to make distributed solutions increasingly more intelligent, higher level agents required deployment of more and more knowledge. This is why semantics has been introduced and ontology knowledge structures shared by agents became an obvious vehicle to reduce the communication traffic and to make the agents more intelligent. In some cases, the ontology converged to the WWW technology (or was combined with it). The direct communication and interaction with and among the devices (not only with their SW modules as virtual representation) became necessary to get faster access to physical devices, to the physical world. The Internet of Things appeared.

This development led to the new vision of the Factory of the Future formulated in the German governmental initiative Industry 4.0, 2013. This vision is nothing else than the extension of the trends in the field of distributed intelligent control combined with new business models supported by accelerated development in the domain of computing and communication. Industry 4.0 is based on the following principles:

- Integration of both the physical and virtual worlds using the Internet of Things and Internet of Services.
- Vertical Integration along the enterprise axis, which means integration of all the information and knowledge-based systems in a company, starting with the real-time control level of shop floor up to the ERP and managerial systems on the top.
- Horizontal Integration along the value chain axis, which means integration of all business activities starting from the supply chain on and up to the product delivery phase (from suppliers to customers).
- Engineering Activities Integration along the life cycle axis from rough idea via design, development, verification, production and testing up to product-lifecycle management (from design to support).

The visions of the three integration axes are based on the following MAS principles: cooperation of distributed autonomous units, ontology knowledge sharing and big data analytics. Industry 4.0 solutions are more and more linked or even coupled with the higher level information systems of the company. Their implementations are influenced by the latest trends in SW engineering exploring service-oriented architectures (SOA). The MAS technology remains to represent an excellent and promising theoretical background for developing an Industry 4.0 solution. The MAS theory can be used to support research activities, to bring new features to these solution explorations, e.g. AI principles, machine learning, data mining and data analytics in general. But the implementations do explore—as a rule—the SOA approaches in broader and broader scale. These are critically simplifying real-life solutions.

This volume of the SOHOMA'15 contributions displays the current trends in intelligent manufacturing, namely the duality of MAS and SOA approaches. It confirms that additional techniques like ontology knowledge structures, machine learning, etc. represent very important and promising topics for further research; they are expected to enrich the current solutions and bring Industry 4.0 visions to industrial practice. In addition, this volume documents many successful research results in this direction.

December 2015   Vladimír Mařík

Preface

This volume gathers the peer reviewed papers which were presented at the fifth edition of the International Workshop "Service Orientation in Holonic and Multi-agent Manufacturing—SOHOMA'15" organized on 5–6 November 2015 by the Institute for Manufacturing (IfM) of the University of Cambridge, UK in collaboration with the CIMR Research Centre in Computer Integrated Manufacturing and Robotics of the University Politehnica of Bucharest, Romania, the LAMIH Laboratory of Industrial and Human Automation Control, Mechanical Engineering and Computer Science of the University of Valenciennes and Hainaut-Cambrésis, France and the CRAN Research Centre for Automatic Control, Nancy of the University of Lorraine, France.

SOHOMA scientific events have been organized since 2011 in the framework of the European project ERRIC, managed by faculty of Automatic Control and Computer Science within the University Politehnica of Bucharest.

The book is structured in seven parts, each one grouping a number of chapters describing research in actual domains of the digital transformation in manufacturing and trends in future manufacturing control: Part I: Applications of Intelligent Products, Part II: Recent Advances in Control of Physical Internet and Interconnected Logistics, Part III: Sustainability Issues in Intelligent Manufacturing Systems, Part IV: Holonic and Multi-Agent System Design for Industry and Services, Part V: Service Oriented Enterprise Management and Control, Part VI: Cloud and Computing-Oriented Manufacturing, Part VII: Smart Grids and Wireless Sensor Networks.

These seven evolution lines have in common concepts, methodologies and implementing solutions for the *Digital Transformation of Manufacturing* (DTM).

The Digital Transformation of Manufacturing is the actual vision and initiative about developing the overall architecture and core technologies to establish a comprehensive, Internet-scale platform for networked production that will encapsulate the right abstractions to link effectively and scalably the various stakeholders (product firms, manufacturing plants, material and component providers, technology and key services providers) to enable the emergence of a feasible and sustainable Internet economy for industrial production.

For the manufacturing domain, the digital transformation is based on the following:

1. **Instrumenting** manufacturing resources (machines, robots, AGVs, ASRSs, products carriers, buffers, a.o.) and environment (workplaces, material flow, tooling, a.o.) which allows: product traceability, production tracking, evaluation of resources' status and quality of services, preventive maintenance...
2. **Interconnecting** orders, products/components/materials, resources in *a service-oriented* approach using multiple communication technologies: wireless, broadband Internet, mobile applications.
3. **Intelligent, distributed control** of production by:

   - *New controls* based on ICT convergence in automation, robotics, vision, multi-agent control, holonic organization; the new controls enable the **smart factory**.
   - *New operations* based on product- and process modelling and simulation. **Ontologies** are used as a "common vocabulary" to provide semantic descriptions/abstract models of the manufacturing domain: *core ontology*— modelling of assembly processes (resources, jobs, dependencies, a.o.); *scene ontology*—modelling flow of products; *events ontology*—modelling various expected/unexpected events and disruptions; these models and knowledge representation enable the **digital factory**.
   - *Novel management* of complex manufacturing value chains (production, supply, sales, delivery, etc.) for networked, **virtual factories**: (a) across manufacturing sites: logistics, material flows; (b) across the product life cycle.

Research in the domain of DTM is determined by the last decades' trend in the goods market towards highly customized products and shorter product life cycles. Such trend is expected to rise in the near future, forcing thus companies to an exhaustive search for achieving responsiveness, flexibility, reduction of costs and increased productivity in their production systems, in order to stay competitive in such new and constantly changing environment. In addition, there is a shift from pure goods dominant logic to service dominant logic which led to service orientation in manufacturing and orienting the design, execution and utilization of the physical product as vehicle for delivering generic or specific services related to that product (in "Product-Service Systems").

How this new vision on digital transformation of manufacturing is achieved? Reaching the above objectives require solutions providing:

- Dynamic reconfigurability of production (re-assigning resource teams, re-planning batches, rescheduling processes) to allow "agile business" in manufacturing;
- Robustness at technical disturbances;

- Efficient execution of production (in terms of cost, productivity, balanced usage of resources);
- Sustainability of manufacturing (proper asset management, controlled power consumption, quality assurance);
- Integration of manufacturing enterprise processes:
  - Vertical integration of the business, MES and shop-floor layers of the manufacturing enterprise;
  - Horizontal integration through value networks.

The solutions adopted for achieving digital transformation of manufacturing are as follows:

A. **Distributed Intelligent Control** at manufacturing execution system (MES) and shop-floor levels, based on ICT frameworks: control distributed over autonomous intelligent units (agents), multi-agent systems (MAS), holonic organization of manufacturing.
B. **Service-Oriented Architecture** (SOA), more and more used as an implementation mean for MAS. SOA represents a technical architecture, a business modelling concept, an integration source and a new way of viewing units of automation within the enterprise. Integration of interoperable business and process information systems at enterprise level are feasible by considering the customized product as "active controller" of the enterprise resources—thus providing consistency between the material and informational flows within the production enterprise. Service orientation in the manufacturing domain is not limited to just Web services, or technology and technical infrastructure either; instead, it reflects a new way of thinking about processes and resources that reinforce the value of commoditization, reuse, semantics and information, and create business value.
C. **Manufacturing Service Bus** (MSB 2.0) integration model: an adaptation of the enterprise service bus (ESB) technology for manufacturing enterprises; it introduces the principle of bus communication between the manufacturing layers acting an intermediary for data flows and assures loose coupling of manufacturing modules.

New developments are induced by digital transformation of manufacturing; they are described in this book:

Cloud manufacturing (CMfg), one of these new lines, has the potential to move from production-oriented manufacturing processes to customer- and service-oriented manufacturing process networks, e.g. by modelling single manufacturing assets as services in a similar way as SaaS or PaaS software service solutions. The cloud manufacturing paradigm moves the intelligent manufacturing system (IMS) vision one step further since it provides service-oriented networked product development models in which service consumers are enabled to configure, select and use customized product realization resources and services, ranging from computer-aided engineering software to reconfigurable manufacturing systems.

   To achieve high levels of productivity growth and agility to market changes, manufacturers will need to leverage *Big Data* sets to drive efficiency across the networked enterprise. There is need for a framework allowing the development of *Manufacturing Cyber Physical Systems* (MCPS) that include capabilities for complex event processing and big data analytics, which are expected to move the manufacturing domain closer towards digital- and cloud manufacturing within the *Contextual Enterprise*.

   A brief description of the book chapters follows.

   Part I reports recent advances and ongoing research in developing *Applications of Intelligent Products*. The intelligent product (IP) model was introduced as a means of motivating supply chains in which products or orders were central as opposed to the organizations that stored or delivered them. This notion of a physical product influencing its own movement through the supply chain is enabled by the evolution of low-cost RFID systems which promise low-cost connection between physical goods and networked information environments. The characteristics of an IP and the fundamental ideas behind it can also be found in other emerging technological topics, such as smart objects, objects in autonomous logistics and the Internet of Things. In manufacturing, the intelligent product is the driver for heterarchical operations scheduling and resource allocation. The IP is one member of the set of Active Order Holons, which together compose the delegate MAS performing collaborative decisions concerning the product's route. This solution allows implementing "product-driven automation" in a completely decentralized mode. This section includes papers describing how the IP concept is used in: hybrid control of radiopharmaceuticals production, improving productivity of construction projects, automation of repair of appliances and necessary information requirements, and End-of-Life information management for a circular economy.

   Part II groups papers devoted to *Physical Internet Simulation, Modelling and Control*. The current instrumenting and interconnecting facilities and the availability of individual information in open-loop supply chains enable new organizations like Physical Internet (PI). One of the key concepts of the PI relies on using standardized containers that are the fundamental unit loads. Physical goods are not directly manipulated by the PI but are encapsulated in standardized containers, called PI-containers. The PI relies on a distributed multi-segment intermodal network. By analogy with the Digital Internet transmitting data packets rather than information/files, the PI-containers constitute the material flow among the different nodes of the PI network. The design of cross-docking hub (in analogy with digital internet, can be seen as a router), allowing the quick, flexible and synchronized transfer of the PI-containers, is essential for the successful development of the Physical Internet. Different types of hubs, denoted PI-hubs, are considered (e.g. road to rail, road to road, ship to rail). The aim of the innovative PI concept is to solve unsustainability present in current supply chains and logistics systems. Papers discuss: automated handling, storage, routing and traceability in the PI context combining spontaneous networking offered by WSN with container virtualization; frameworks for instrumenting PI in a collective of "smart" PI-containers in interaction; crowdsourcing solutions to last mile delivery in e-commerce environment; open tracing container and IoT in automotive and transport fields.

Part III analyses *Sustainability Issues in Intelligent Manufacturing Systems.* Two perspectives are considered: (1) The needs to add sustainability to efficiency performance in IMS design, and (2) Approaching these needs using concepts from IMS engineering methods in the context of sustainable manufacturing systems design. Directions indicated by the reported research: (1) Go-green holons as green artefact that help the system designer to implement solutions for sustainable IMS, and (2) Defining a set of guidelines that enforce system engineers to think about their main designs choices of the sustainable parameters to be taken into account in the new type of IMS. Key requirements for resilient production systems are also developed by establishing the links between production disruption and the required resilient control and tracking capabilities in production systems. A semantic model of requirements in large and complex manufacturing systems, based on business concepts and modelled with resource description framework (RDF), is included in this book section. Related to sustainability, Part III also analyses how the human operator is integrated in the IMS's control architecture. "Human-in-the-loop" Intelligent Manufacturing Control Systems consider the intervention of humans (typically, information providing, decision-making or direct action on physical components) during the intelligent control of any functions relevant to the operational level of manufacturing, such as scheduling, maintenance, monitoring, supply, etc. With a human-centred design approach in IMS, human resources can be assisted by recent ICT tools helping them detecting in advance problems, propose efficient solutions and take decision and action.

Part IV reports recent advances in *Holonic and Multi-Agent System Design for Industry and Services.* Nowadays, industry is seeking for models and methods that are not only able to provide efficient global batch production performance, but also reactively facing a growing set of unpredicted events. One important research activity in the field focuses on holonic and multi-agent control systems that integrate predictive, proactive and reactive mechanisms into agents/holons. The holonic approach is the main engine for the digital transformation of manufacturing at manufacturing execution system (MES) middle layer and shop-floor production control layer level in what concerns "Distribution" and "Intelligence". The holonic manufacturing paradigm is based on defining a main set of assets: resources (technology, humans—reflecting the producer's profile, capabilities, skills), orders (reflecting the business solutions) and products (reflecting the client's needs, value propositions)—represented by holons communicating and collaborating in holarchies to reach a common goal—expressed by orders. As

$$[Holon] \leftarrow [Physical\ Asset] + [Agent = Information\ counterpart],$$

it becomes possible to solve at informational level all specific activities of the physical layer: mixed batch planning, product scheduling, resource allocation, inventory update, product routing, execution, packaging, tracking and quality control:

- Triggered by real-time events gathered from the manufacturing structure processes and devices;
- Controlled in real time, with orchestration and choreography assured by SOA in standard, secure mode.

Thus, the holarchy created by the holons defined for any holonic manufacturing system (HMS) acts as a "Physical Multi-agent System—PMAS", transposing in the physical realm the inherent distribution induced by an agent implementation framework, according to defined manufacturing ontologies.

The demand for production systems running in complex and disturbed environments requires considering new paradigms and technologies that provide flexibility, robustness, agility and responsiveness. Holonic systems are by definition targeting challenges that include coping with the heterogeneous nature of industrial systems and their online interactive nature in combination with competitive pressures. Multi-agent systems is a suitable approach to address these challenges by offering an alternative way to design control systems, based on the decentralization of control functions over distributed autonomous and cooperative entities. Chapters of this Part IV describe: coordinating mechanisms in HMS; automatic diagnostic methods to increase dependability by using model-checking at runtime; interfacing BDI agent systems with geometric reasoning in robotized manufacturing; nervousness control mechanisms for semi-heterarchical MES; and applications of HMS in industry.

Part V groups papers dealing with *Service Oriented Enterprise Management and Control*. Integrating the concepts of services into HMS gives rise to a new type of systems: service-oriented holonic manufacturing systems (SoHMS). SoHMS is underpinned by the use of a structure based on repeatability and reusability of manufacturing operations. Process families are formed by a collection of process modules representing manufacturing operations. By adopting the principles of SOA into HMS, such manufacturing operations can be standardized into manufacturing services (MServices) possessing a proper identification and description. Thus, the service becomes the main element of negotiation and exchange among holons. Conceiving manufacturing services and manufacturing processes specifications allows the HMS's control architecture to explore manufacturing flexibility at process level with the decomposition and encapsulation of processes.

Following the IT-based approach which defines a service as "a single activity or a series of activities of a more or less intangible nature that normally takes place in the interactions between client and service provider, which is offered as a solution to achieve desired end results for the client" in a SoHMS manufacturing operations can be represented by MServices that are executed over a product and can be realized by one or several resources in the system. MServices, as they represent validated operations, can be readily available to integrate different production processes, thus bringing the benefit of *reusability*. Moreover, resource capabilities are determined by the collection of MServices it offers. This facilitates the *integration of legacy systems* and different vendor technologies, as MService descriptions are determined according to their nature, in terms of added transformations, with no regard of the methods that are used for their application. This allows a *complete separation of process specification from the knowledge on the production floor* making it implementable in any SoHMS platform providing the necessary MServices with the same application service ontology.

Service orientation is emerging at multiple organizational levels in enterprise business, and leverages technology in response to the growing need for greater business integration, flexibility and agility of manufacturing enterprises. Closely related to IT infrastructures of Web services, the service-oriented enterprise architecture represents a technical architecture, a business modelling concept, an integration source and a new way of viewing units of control within the enterprise. Business and process information systems integration and interoperability are feasible by *considering the customized product* as "*active controller*" of the enterprise resources—thus providing consistency between material and informational flows. The areas of service-oriented computing and multi-agent systems are getting closer, trying to deal with the same kind of environments formed by loose-coupled, flexible, persistent and distributed tasks. An example is the new approach of service-oriented multi-agent systems (SoMAS).

The unifying approach of the authors' contributions for this Part V of the book relies on the methodology and practice of disaggregating siloed, tightly coupled business, MES and shop-floor processes into loosely coupled services and mapping them to IT services, sequencing, synchronizing and orchestrating their execution. Research is reported in: function block orchestration of services in distributed automation and performance evaluation of Web services; MAS with service-oriented agents for dynamic rescheduling work force tasks during operations; virtual commissioning-based development of a service-oriented holonic control for retrofit manufacturing systems; security solution for service-oriented manufacturing architectures that uses a public-key infrastructure to generate certificates and propagate trust at runtime.

Part VI is devoted to *Cloud and Computing-Oriented Manufacturing*, which represent major trends in modern manufacturing. Cloud manufacturing (CMfg) and MES virtualization were introduced as a networked and service-oriented manufacturing model, focusing on the new opportunities in networked manufacturing area, as enabled by the emergence of cloud computing platforms. The cloud-based service delivery model for the manufacturing industry includes product design, batch planning, product scheduling, real-time manufacturing control, testing, management and all other stages of a product's life cycle.

CMfg derives not only from cloud computing, but also from related concepts and technologies such as the Internet of Things—IoT (core enabling technology for goods tracking and product-centric control), 3D modelling and printing (core enabling technology for digital manufacturing). In CMfg applications, various manufacturing resources and abilities can be intelligently sensed and connected into a wider Internet, and automatically managed and controlled using both (either) IoT and (or) cloud solutions. The key difference between cloud computing and CMfg is that resources involved in cloud computing are primarily computational (e.g. server, storage, network, software), while in CMfg all manufacturing resources and abilities involved in the whole life cycle of manufacturing are aimed to be provided for the user in different service models.

Papers in this section present resource virtualization techniques and resource sharing in manufacturing environments. Resources and resource capabilities virtualization and modelling represent the starting point for manufacturing services encapsulation in the cloud. There is also shown that CMfg is clearly an applicable business model for 3D-printing—a novel direct digital manufacturing technology. In cyber-physical system (CPS) approach of manufacturing, a major challenge is to integrate the computational decisional components (i.e. cyber part) with the physical automation systems and devices (i.e. physical part) to create such network of smart cyber-physical components at MES and shop-floor levels. Some works present the development of standardized interfaces for HMES that can be used to access physical automation components by the cyber layer in CPS. A chapter of this section investigates the software-defined networking (SDN) concept adoption for the manufacturing product design and operational flow, by promoting the logical-only centralization of the shop-floor operations control within the manufacturing shared-cloud for clusters of manufacturing networks.

Part VII gathers contributions in the field of *Smart Grids and Wireless Sensor Networks* management and control with multi-agent implementing. Technological advances in wireless sensor networks are enabling new levels of distributed intelligence in several forms such as "active products" that interact with the working environment and smart metering for monitoring the history of products over their entire life cycle and the status and performances of resources. These distributed intelligences offer new opportunities for reducing myopic decision-making in manufacturing control systems, thereby potentially enhancing their sustainability.

Design of such MAS frameworks for distributed intelligent control and development of applications integrating intelligent-embedded devices are reported in Part VII for several representative domains. Thus, a solution for space system management is proposed by creating a self-organizing team of intelligent agents, associated to spacecraft modules, conducting negotiations and capable of both planning their behaviour individually in real time and working in groups in order to ensure coordinated decisions. Then, an embedded multi-agent system for managing sink nodes and clusters of wireless sensor networks is proposed and finally demonstrated in an oil and gas refinery application. To reduce the communication overhead in the MAS, the wireless sensor network is clustered which leads to a hierarchical structure for the WSN composed of two types of sensor nodes: sink nodes (cluster heads) and anchor nodes (sending sensory data to the sink nodes) allowing for data aggregation. Finally, this section describes a methodology and framework for the development of new control architectures based on uncertainty management and self-reconfigurability of smart power grids.

The book offers a new integrated vision on *complexity, Big Data* and *virtualization in Computing-Oriented Manufacturing*, combining emergent information and communication technologies, control with distributed intelligence and MAS implementation and total enterprise integration solutions running in truly distributed and ubiquitous environments. The IMS philosophy adopts heterarchical and collaborative control as its information system architecture. The behaviour of the entire manufacturing system therefore becomes collaborative, determined by many interacting subsystems that may have their own independent interests, values and modes of operation. Also, the enrichment of distributed systems with biology-inspired mechanisms supports dynamic structure reconfiguration, thus handling more effectively condition changes and unexpected disturbances, and minimizing their effects.

All these aspects are treated in the present book, which we hope you will find useful reading.

December 2015

Theodor Borangiu
Damien Trentesaux
André Thomas
Duncan McFarlane

Contents

Part I
Applications of Intelligent Products

Centralized HMES with Environment Adaptation for Production of Radiopharmaceuticals

Silviu Răileanu, Theodor Borangiu and Andrei Silişteanu

Abstract The paper presents an intelligent hybrid control solution for the production of radiopharmaceuticals that irradiates radioactive isotopes of neutron-defficient radionuclides type by in cyclotrons. To achieve requirements such as: highest number of orders accepted daily, shortest production time and safe operating conditions, a hybrid control system based on a dual architecture: centralized HMES with ILOG planning and decentralized parameter monitoring and control via SCADA is developed. Experimental results are reported.

Keywords Holonic manufacturing execution system · SCADA · Environment monitoring · Intelligent product · Constraint programming · Radiopharmaceuticals

1 Introduction. The Radiopharmaceutical Production Process

Radiopharmaceuticals are products that contain radioactive materials called radioisotopes for the treatment of many life threatening diseases (therapeutic radiopharmaceuticals) and established means for medical diagnosis (diagnostic radiopharmaceuticals) and research in many disciplines of life sciences [1, 2]. Radiopharmaceuticals produced using a cyclotron and dedicated radiochemistry equipment and laboratories are used for positron emission tomography (PET) and single photon emission computed tomography (SPECT) [3].

S. Răileanu (✉) · T. Borangiu · A. Silişteanu
Department of Automation and Industrial Informatics,
University Politehnica of Bucharest, 060042 Bucureşti, Romania
e-mail: silviu.raileanu@cimr.pub.ro

T. Borangiu
e-mail: theodor.borangiu@cimr.pub.ro

A. Silişteanu
e-mail: andrei.silisteanu@cimr.pub.ro

© Springer International Publishing Switzerland 2016
T. Borangiu et al. (eds.), *Service Orientation in Holonic and Multi-Agent Manufacturing*, Studies in Computational Intelligence 640,
DOI 10.1007/978-3-319-30337-6_1

The production of radiopharmaceuticals needs performing two processes: (1) Obtaining the radionuclides on which the pharmaceutical is based, i.e. the radioactive isotopes of elements with atomic numbers less than that of bismuth; (2) Preparing and packaging of complete radiopharmaceuticals. Proton-deficient radionuclides represent the first category of radioactive isotopes of elements (e.g. technetium-99m) and are produced in a nuclear reactor [4], whereas neutron-deficient radionuclides (those with fewer neutrons in the nucleus than those required for stability) are the second category of radioactive isotopes of elements being most easily produced using a proton accelerator such as a medical cyclotron [5]. An intelligent control solution for the production of radiopharmaceuticals that uses the second type of radioactive isotope will be proposed in this paper.

The main challenges of a cyclotron-based radiopharmaceutical production line (composed of: facilities, control and supervising system) are to manufacture valid nuclear medicine products in the shortest possible time, safely for the employees and surrounding environment. Such a production line is specialized in producing small batches of products in small volumes, according to the orders received from hospitals and PET centres. While having a specific chemical structure, radioactivity and usage, each product follows the same manufacturing path: radio-isotopes are produced in a particle accelerator (cyclotron), then transferred into technology isolators for chemical synthesis followed by portioning (vial dispensing) the bulk product, and quality control of the final product by conformity tests on multiple parameters; in the last stage, valid products are packed and transported to clients in shielded containers. The manufacturing stages (S1)–(S4) for the cyclotron-based production line of radiopharmaceuticals are presented in Fig. 1, and involve one functional block (coloured in light grey) for each stage; a fifth stage (S5) should be considered for the transport of the final, valid products to the clients (e.g., hospitals, PET centres).

Normally, due to its special reliability provided by the manufacturer, and also to its high acquisition and maintenance costs, production lines of this type have only one cyclotron resource [6, 7] for stage 1—raw materials irradiation. On the other hand for single points of failure avoidance and fault tolerance reasons, stage 2 (product configuring) and stage 3 (portioning of the bulk product) may use respectively any of two identical resources (synthesis modules 1, 2 and robotized dispenser 1, 2), capable to replace each other in the event of a breakdown or

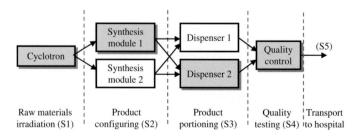

Fig. 1 Manufacturing stages for radiopharmaceutical products

maintenance work. The radiopharmaceutical production process ends with quality control of the final product (S4) consisting of 5 tests, each being done on a different, dedicated laboratory equipment.

The three production resources together with the quality testing equipment are disposed in *flow shop* processing mode: products flow in one single direction [8]. Raw materials enter the cyclotron and are irradiated, the output being fed to one synthesis module where chemical reactions occur; the resulting bulk radiopharmaceutical product is then portioned and eventually diluted in one robotized dispenser box; samples of vials are sent to the quality control room for multi-parameter tests; finally, the vials are packed and labelled for transport to the client. In this manufacturing mode, the product in different stages passes through capillary tubes from one resource to the next and receives a service/operation. The mode of processing is *non-preemptive*, and the fixed set of precedence constraints among operations is defined.

The constraint $o_{i-1} \prec o_j, o_i = o(S_i), 1 \leq i \leq 4$ means that the processing of operation o_{i-1} must be completed before o_j can be started, i.e., the set of 4 operations carried out in stages (S1)–(S4) is *totally ordered* by the operator $\prec$. This set of operations on irradiated materials ordered by the precedence relation can be represented as a digraph in which nodes correspond to operations and arcs to precedence constraints.

Basically, production planning in flow shop material processing mode with partially duplicated resources must provide a balanced usage of resources of the same type considering the scheduled maintenance periods; for the radiopharmaceuticals production line based on the cyclotron as proton accelerator, the doubled resources in stages 2 (synthesis modules 1, 2) and 3 (robotized dispenser 1, 2) are scheduled for periodical maintenance every 6 month, with a 3 month offset of resources 2 relative to resources 1 (see Fig. 2). This allows normal usage of one series of resources (e.g., 1 or 2) for the first 3 months after maintenance followed by replacement with the other series of resources (e.g., 2 or 1) for the next 3 months.

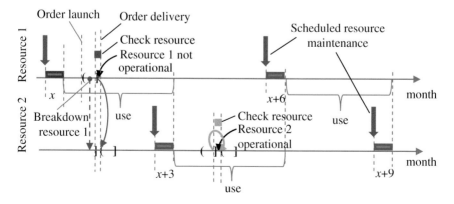

Fig. 2 Utilization of duplicated resources in stages 2 (synthesis modules 1, 2) and 3 (robotized dispenser 1, 2) during maintenance periods and at breakdown events

Starting from the deadlines for product delivery defined by clients (hospitals), there will be computed the daily moments of time when orders should be launched for production execution (the time intervals "( ]" in the timing of Fig. 2), considering additionally: (a) the estimated transportation time of the final product to the client, and (b) the testing period of time following the normal completion of any production process, when all resources are checked whether they are operational. If, during this test, one resource is found not operational, it will be substituted with its stand-in one (e.g. resource 1 replaced with resource 2, as depicted in Fig. 2). A similar replacement is possible for the robotized dispensers in real time at resource breakdown during stage 3 (see the "breakdown resource 1" event represented in Fig. 2) [6].

Due to the specificity of processes transforming and handling radioactive materials and products, the basic functions of the global production control system are: service orientation and monitoring continuously the parameters of: (1) manufacturing processes and (2) resources, and (3) environment parameters of production rooms [9, 10].

2 Control Architecture with Centralized HMES and SCADA for Distributed Parameter Monitoring

In order to achieve the requirements stated above (maximum number of accepted orders, shortest possible production time and safe operating conditions) a hybrid control system is proposed for radiopharmaceuticals production based on a dual architecture: centralized HMES (Holonic Manufacturing Execution System) and decentralized parameter monitoring and control via SCADA.

2.1 Layered Hybrid Control Architecture: HMES–SCADA

Figure 3 shows the hybrid control architecture proposed for the manufacturing system producing radiopharmaceuticals in shop floor processing mode. The topology of the control architecture is multi-layered: (a) Manufacturing Execution System (MES) layer—planning and resource allocation, data storage and reports generation; (b) application layer with SCADA—parameter monitoring and adjusting; (c) resource control layer.

The global control is exerted on these three layers by two subsystems:

1. **Centralized HMES** for: (a) long-term resource allocation balancing the usage time of those replicated for stages 2 and 3 in the context of scheduled maintenance, (b) short-term (24 h) production planning optimizing a cost function (i.e. minimize: manufacturing time/raw material waste, etc.), (c) manage the centralized data storage for executed product batches and generate reports.

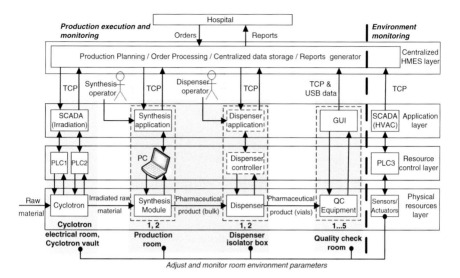

Fig. 3 The hybrid, multi-layer control architecture of radiopharmaceuticals production line

2. **Decentralized and reactive SCADA** for: (a) monitoring and adjusting process and environment (production rooms) parameters in response to unforeseen disturbances in order to deliver requested, valid products at agreed deadlines, (b) detecting resource failures and initiating the replacement process upon receiving the authorization from HMES, and (c) collecting data during manufacturing for product traceability and description according to IP (Intelligent Product) conventions [11].

2.2   The Holonic Control Mode

The proposed control model is based on the PROSA and ADACOR reference architectures and is customized for the flow shop layout and operating mode, in which the product recipe is configured directly by the client (Make To Order production) and embedded into the production order [12, 13]. Another specific feature of the control model is that the environment parameters such as pressure, temperature, humidity, number of particles and radioactivity levels strongly affect the production process. This is why the control model is extended with an entity— the Environment Holon—that models the process rooms (cyclotron electrical room, cyclotron vault, production room, dispenser technical isolator box, and quality testing room) together with the instrumentation that measures and adjusts these parameters.

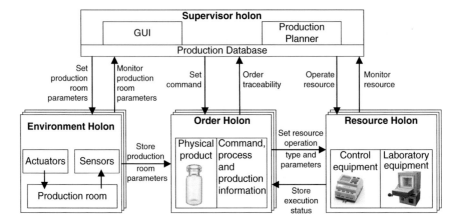

Fig. 4 Entities, functions and interactions in the proposed holonic production control system

The structure of the control model is composed of the following entities (Fig. 4). The **Supervisor Holon** (SH) is responsible with:

- *Optimizing production*: (a) resource allocation subject to (fixed) scheduled maintenance periods, alternative resource re assigning at breakdown (while execution in stage 3) or not operational status (detected at resources checking post execution of stage 3), and (b) operations planning subject to:

  - Minimizing the duration of production execution considering required quantities of radiopharmaceutical products and imposed/agreed delivery times;.
  - Maximizing and balancing the utilization of resources subject to maintenance periods and failure events;
  - On line adaptation to variations of environmental parameters.

- *Setting orders* (associate hospital order with physical product); configure process and production room parameters for planned production and assigned resources;
- *Authorizing the adjustment of process rooms parameters* (e.g., pressure, temperature, relative humidity) upon request of the Environmental Holon;
- *Centralizing data* about resource status, process rooms parameters and product execution; *storing production logs/history* files into a centralized and replicated Production Database;
- *Generating traceability reports* and *IP descriptions* of radiopharmaceuticals;
- *Keeping BOM* (Bill Of Materials) updated.

The **Resource Holons** (RHs) encapsulate both the decisional/informational part and the associated physical resources used in the production line: cyclotron, synthesis unit, dispenser and quality test (laboratory equipment). Each physical resource is controlled as an independent automation island, and the objective of the proposed control architecture is to integrate in a uniform manner these islands (modelled as RHs) by help of the SH for Order Holon (OH) execution. In order to facilitate resource integration, the following attributes and methods were proposed to characterize an RH:

1. Attributes (information about system resources):

    – string: *Resource identification* (name, ID)—the name and identifier of the current resource;
    – string: *Resource description*;
    – int: *Number of operations* (nr_op_res)—the number of operations the current resource is capable of executing;
    – {string}$_i$, $i = 1...$nr_op_res: *Operations names*—the set of operations the current resource is capable of executing;
    – {int}$_i$, $i = 1...$nr_op_res: *Execution time* [seconds]—the set of execution times of the operations that can be done on the current resource;
    – string: *Current operation*—the operation the current resource is executing. This is one of the operations the resource is capable of executing;
    – int: *Maximum delay* [seconds]—the amount of time the product can be additionally held in the current stage (Si) due to problems with resources or environment parameters at the next production stage (Si + 1);
    – int: *Energy consumption* [Wh]—the energy consumed at the current stage;
    – {string}$_i$, $i = 1...$nr_op_res: *Input*—the description of the current stage's input for operation i;
    – {string}$_i$, $i = 1...$nr_op_res: *Output*—the description of the current stage's output for operation i;
    – date: *Online time*—the time and date starting from which the resource is online;
    – time: *Idle time*—the interval of time during which the resource was unused;
    – time: *Working time*—the interval of time during which the resource was used.

2. Methods to access the resource (types of messages the current resource responds):

    – void: *Configure operation* (target resource, parameter index, parameter value)—sets a specified parameter of the targeted resource to an imposed value;
    – void: *Start operation* (target resource, operation index)—triggers the start of the specified operation on the targeted resource;
    – int: *Request status* (target resource)—returns the current status of the targeted resource; the status can be: offline, online and idle, online and working.

An instantiation for the attributes of the available production resources (cyclotron, synthesis module and dispenser) modelled as RHs is detailed in Table 1.

Table 1 Instantiation of manufacturing resources modelled as RHs

Resource	Cyclotron	Synthesis modules	Dispenser
Description	Irradiate two liquid targets simultaneously	Automatically execute a prebuilt synthesis algorithm to produce one of two types of products: FDG and NaF; recover enriched water after synthesis; measure input and output radioactivity of raw material/radiopharmaceutical compound (bulk solution)	Sterile dispense radiopharmaceutical compound into vials with final product using a robotic arm inside a technical isolator; fill, cap and crimp final product into vials and then into shielded containers to minimize the exposure of the operators; create dispensing recipe according to orders; read bar codes; measure initial and final product activity
Operation	Irradiation	Synthesis	Dispensing
Execution time	Irradiation time: 1–2 h	Approx. 23 min (FDG) Approx. 7 min (NaF)	5 min for the 1st vial, then 2 min for each next vial
Maximum delay	1 h (FDG half-life is 1 h and 9 min)	30 min (due to increased dust particles in the dispenser)	15 min
Energy	100 KWh	300 Wh	400 Wh
Input	Enriched water	Irradiated enriched water from cyclotron	Radiopharmaceutical compound (bulk solution)
Output	Irradiated enriched water	FDG/NaF radiopharmaceutical compound	Final radiopharmaceutical product (delivered)

The **Order Holon** (OH) holds all the information needed to fulfil a command starting from the specifications of the hospital (client), production information used for resource and environment parameterization and accompanying information generated once the product leaves each production stage. The OH is an aggregate entity consisting of: (1) the information needed for demand identification (product type, radioactivity level, hospital #ID, and delivery time), execution of processes (sequence of resources, operations and process parameters as resulted from production planning) and traceability (sequence of resources, operations, execution reports as resulted from the physical process) and (2) the physical product. From the OH lifecycle depicted in Fig. 5 it can be seen that the physical product results at the termination of stage 4 when the physical-informational association is performed.

The informational part of an OH for the different stages of its lifecycle is given below (see Fig. 5); a generic structure was sought:

//Information about the client (from "Hospital command" in Fig. 5):

– string: *Order identification* (name, ID)—name and identifier of the current OH;
– int: *Product type*—name of the pharmaceutical product that is being realized;
– int: *Radioactivity level*—radioactivity level needed for the product;

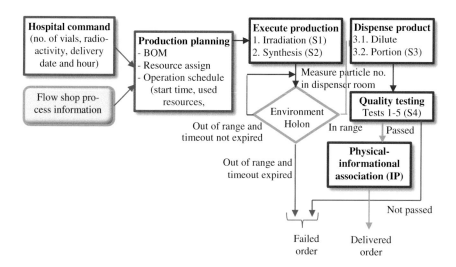

Fig. 5 Order Holon lifecycle

– string: *Hospital*—name of the hospital (client);
– time: *Delivery time*—hour at which the product must be delivered the next day.

//Information about the production process ("Flow shop process information" in Fig. 5):

– int: *Number of operations* (nr_op_order)—number of operations the current order must receive;
– {string}$_i$, $i = 1…$nr_op_order: *Operations names*—set of operations the current OH must receive;
– {string}$_i$, $i = 1…$nr_op_order: Precedencies—operations that must precede operation i.

//Information computed in the planning process ("Production planning" in Fig. 5):

– {int}$_i$, $i = 1…$nr_op_order: *Resources*—the set of resources that must be visited in order to execute the operations stated above (Operations names);
– {int}$_i$, $i = 1…$nr_op_order: *Parameters*—the set of parameters that configure the resources in order to receive the desired operations;
– {int}$_i$, $i = 1…$nr_op_order: *Maximum execution time*—the time needed to execute the operation "Operations names i" on the resource "Resources i".

//Information gathered during production process ((S1), (S2), (S3) and Measurement of particle no. in Fig. 5):

– {int}$_i$, $i = 1…$nr_op_order: *Timing*—the time spent to perform operation i;
– {string}$_i$, $i = 1…$nr_op_order: *Parameters*—parameters used by resource i to execute operation i;

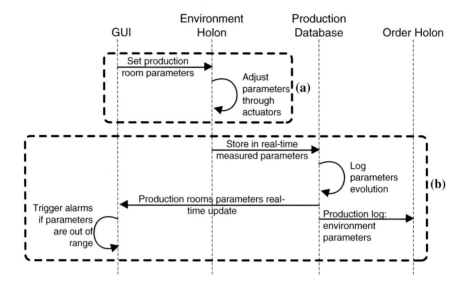

Fig. 6 Real-time operating mode of EH for parameter monitoring and adjusting, and data log

- int: *Quantity*—the quantity of the product that will be delivered to the hospital;
- int: *RFID*—the code which associates the physical product with the current information ("IP" in Fig. 5).

The **Environment Holon** (EH) checks if the process and environment parameters are in range, validates the operations executed by RHs and triggers alarms when radioactivity levels exceed normal values or when the evolution of other parameters endangers production or human security. Figure 6 illustrates the operating mode of the EH.

The parameters monitored by the EH are:

- **Pressure**: in cyclotron vault, production room and dispenser isolator box;
- **Temperature**: in cyclotron electrical room, production room and dispenser isolator box;
- **Humidity**: in cyclotron electrical room, production room and dispenser room;
- **Number of particles** in the dispenser isolator box: if the number of particles is not in range (above product safety threshold) the process waits for a maximum timeout of 30 min to allow this parameter to re-enter in range. If it re-enters the range the process continues and dispensing (dilution and portioning) is delayed with the corresponding amount of time—which also delays delivery; otherwise production is abandoned and the production order fails;
- **Radioactivity level**: in production room, control room and dispenser room.

The operation of the EH is materialized through intelligent actuators and sensors integrated in SCADA, to which a software agent is associated to interface these devices with the centralized HMES. The Environment Holon adjusts process room parameters (pressure, temperature, humidity, number of particles) as imposed by the Supervisor Holon through GUI (Fig. 6a) and sends information back concerning production monitoring, logging and history, adjusts production planning in real-time (online run) and triggers alarms in case parameters are out of range (Fig. 6b).

3 Optimized Production Planning

The objective of the production planning process is to optimize the execution of radiopharmaceutical products in order to undertake as much demands as possible while minimizing the production time and respecting the constraints imposed by the physical installation, the environment and by the client (e.g., hospital).

In the production planning process the following terms will be used: *demand* (one vial needed by a hospital), *command* (a set of demands that are produced together and contain the same type of product) and *batch* (the set of all demands that must arrive at the same hospital; a batch can contain different products). In this context the optimization problem is described by:

- An **input set of data** representing the demands which are characterized by (product type, requested activity, client, delivery date):

  - {*demands*} = {(product type, requested activity, hospital, delivery time, requested volume)$_i$, $i = 1…n$, n being the number of vials requested for the current day}.

- The **decision variables** describing:

  - How the individual demands are allocated to commands:
    {*where*} = {(command)$_{index}$, command = $1…m$, *index* = $1…n$, m being the maximum number of commands that are processed within a day, n being the total number of vials requested for the current day};
  - How are commands processed:
    {*commands*} = {(starting time, maximum irradiation, quantity of enriched water, product type)$_j$, $j = 1…m$, m being the maximum number of commands that are processed within a day}.
  - If individual demands are processed:
    {*processed*} = {(true/false)$_{index}$, *index* = $1…n$, indicating whether demand i is processed or not due to invalid constraints such as tight delivery time}.

- The **constraints**:

  - The sum of all volumes of the demands within a command should be less than the volume of the irradiated target:

$$\sum_{i=1}^{n} V_i * \frac{desired\ activity_i}{maximum\ activity} \leq target\ volume - loss$$

where: demand i will be produced within the current command, V_i is the requested volume of demand i, "desired activity$_i$" is the activity of the product in demand i, "maximum activity" is the activity at which is irradiated the raw material for the current command, "target volume" is the maximum raw material quantity that can be irradiated in a production cycle, and "loss" is the quantity of product that is lost when transporting from one stage to another.

- Each demand should have the required reactivity at the requested delivery time for all demands within command k, $1 \leq k \leq m$:

$$starting\ time + \sum_{r=1}^{4} Tr + T5(i) < delivery\ time_i, 1 \leq i \leq n$$

where $T1...T4$ are the maximum delays when manufacturing the current demand (irradiation time, synthesis time, dispensing time, quality testing time) and $T5(i)$ is the maximum transportation time to the related hospital.
- Do not use the production facility over maintenance periods:

$$production\ intervals \cap maintenance\ intervals = \emptyset,$$

where "production intervals" are defined by starting time of the command and its duration and "maintenance intervals" are predefined based on a fixed scheme which takes into account the resources' usage (see Fig. 2).

- Possible **objective functions**:

    - minimize production time (optimization problem) at command level;
    - maximize the number of commands for daily client orders;
    - minimize the quantity of daily lost raw materials.

By analysing the requirements stated above it can be seen that the optimization problem is first of all a matching problem (which demand is allocated to which command) which is subject to a set of constraints (delivery dates of the vials together with a given quantity and radioactivity level). Finally, since there is flexibility when irradiating the raw materials (less material can be irradiated at a higher level, and then diluted when dispensing), an objective function represented by production time can be added.

Thus, the planning problem deals with *combinatorial optimization* and detailed scheduling, both aspects being tacked in literature by Constraint Programming (CP) approaches [14, 15]. A comprehensive list of CP solvers addressing various combinatorial problems can be found at [16]. Since the problem optimization

option must be included as a functionality of the Supervisor Holon (Fig. 4), the
IBM ILOG OPL optimization engine [17] was chosen because it can be easily
integrated with separate applications using standard C++, C# or JAVA interfaces.
This is facilitated through the Concert technology.

The procedure described above was integrated into the radiopharmaceuticals
production process both for *optimized offline planning* (minimize production
duration while respecting imposed deadlines—offline run) and for *online agile
execution* (online adaptation to variations of environment parameters which can
delay commands being executed within the same day—online run).

Offline run

1. Demands are gathered for the next production day.
2. Maintenance restrictions are introduced as constraints into the CP model.
3. The ILOG model is called based on the set of demands for the next day:

   (a) If a feasible solution is reached, the demands are accepted as received and
       the production plan is transmitted to the distributed operating control level
       (Fig. 3) in order to be implemented;
   (b) If no feasible solution is reached (due to conflicting time constraints or tight
       deadlines) new deadlines are proposed based on the maximum load of the
       production system and on the rule "first came first serve" in order to fulfil as
       much as possible demands.

Online run

1. Apply process and environment parameter configuring via SCADA according to
   the off line computed commands.
2. Measure environment parameters which affect production time (dust particles in
   dispenser chamber and radioactivity levels).
3. If parameters are out of range and the current command is delayed, replan the
   next commands taking into account the new constraints. For any command,
   since the maximum allowed delay in production (30 min) is less than the
   maximum delay accepted for delivery (1 h) the worst case scenario is to use the
   off-line computed production plan and just delay it.

ILOG optimization sequence
The following optimization sequence will be run for a maximum amount of raw
material max_irr_vol (capacity of max. irradiated target) processed for any
command:

1. Consider all demands valid for scheduling ($processed(d)$ = true, $d = 1…n$);
2. Order demands based on product type;
3. Choose the highest activity required (all other products will be diluted in order
   to obtain an inferior activity) for each product type (max_irr_level);

4. For all demands with the same product type compute the sum:

$$\text{sum\_prod} = \sum_{\text{for all products of same type}} requested\ volume \cdot \frac{desired\ activity}{\max.\ activity}$$

Clearly, sum_prod ≤ max_irr_target.

5. Based on the requested product quantity (sum_prod), on the maximum activity (max_irr_level) and on the number of vials, an estimated production time (makespan) for the demands is computed, considering $Ti, 1 \leq i \leq 4$

6. Test if the production intervals with the width computed at step 5 can be scheduled one after another, with a break between them of 1:30 h (resource checking periods between successive commands execution, see Fig. 2), without invalidating the delivery times for each demand:

- For all c in commands
- For all d in demands

  - p = the command processed before (c)
  - If p = null (c is the first command to be produced)

    - YES: production starting time(c) = 6:00 (the installation begins to function at 6:00 in the morning)
    - NO: otherwise production starting time(c) = production ending time (p) + 90 min

  - If (delivery_time(d)) < makespan(c) + production break + production starting time(c)

    - NO: Eliminate demand d from the schedule (processed(d) = false) and goto 2

7. Compute the remaining raw material quantity (rem_raw_mat) that can be used with the associated deadline and maximum activity

- For all c in commands
- For all d in demands, processed(d) == true

  - rem_raw_mat(c) = max_irr_volume—sum_prod

8. Test if the demands eliminated can be produced using the remaining raw material quantity computed at step 7

- For all c in commands
- For all d in demands, processed(d) == false

  – If the product of demand d is the same as the product manufactured in command c and there is enough remaining raw material in command c to produce demand d:

    - Propose a new delivery time for demand d
    - Set processed(d) == true
    - Recompute sum_prod for command c with the new considered demand d
    - Goto 7

The output of the optimization sequence will consist of: (1) the demands scheduled for production based on the imposed (step 6) or negotiated (step 8) delivery times and (2) the unfeasible demands which cannot be produced due to the tight deadline. The optimization sequence is run one day before production execution and the processed quantity of raw material processed is limited to the quantity of the irradiation target. Any command above the maximum raw material quantity that can fill the cyclotron's irradiation target will be abandoned.

4 Experimental Results. Conclusions

The ILOG sequence designed in Sect. 3 was tested on a set of 10 demands grouped into two different product categories (FDG with index 1 and NaF with index 2). The execution times are described in Table 1 for each stage: irradiation, synthesis and dispensing. The characteristics of the demands, which represent the input to the optimization algorithm, are described in Table 2.

If produced individually, the starting time of each demand would be computed based on the activity (how much time the raw material stays in the cyclotron, Table 1) and on the product type (what type of synthesis is applied to the irradiated raw material, Table 1). To this amount of time a fixed duration will be added for dispensing (Table 1) and a fixed duration for installation cleaning (1:30 h). The sum between the irradiation time, synthesis time, dispensing and cleaning time is the time needed to execute each demand. The starting time is computed by subtracting the production duration from the delivery time. A theoretical scheduling is given in Fig. 7.

Analysing Fig. 7 it can be seen that there are overlapping cases between the production times of the demands which makes it impossible to execute all of them individually. By applying the optimization procedure, demands are grouped into commands which are executed together using the same irradiated raw material—allowing thus to maximize the number of executed demands. The only constraints

Table 2 Hospitals demands

Index	Activity (MBq)	Product type	Quantity (µL)	Delivery time (minutes from 0:00)	Production duration (computed)	Start time (computed)
1	600	1	500	1080	214	866
2	600	1	500	900	214	686
3	800	1	500	1080	226	854
4	1400	1	500	1140	262	878
5	700	1	500	1170	220	950
6	1400	2	500	780	246	534
7	1500	2	500	840	252	588
8	900	2	500	840	216	624
9	1500	2	500	860	252	608
10	600	2	500	880	198	682

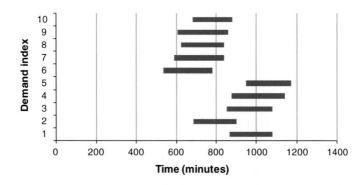

Fig. 7 Gantt chart for demand realization without optimized offline planning

are: a single type of product can be executed at a given time, a maximum of 3.5 ml can be irradiated and if the raw material is irradiated for obtaining the highest activity. This means that less irradiated raw material is used for commands with lower activity. Thus, the advantage of production optimization is that it reduces the production cycles by combining demands into a single command. As can be seen from Fig. 8 the demands can be grouped into 2 separate commands but there is one demand (2, the one marked with red in Table 2) that exceeds the production interval attributed to product 2. This demand cannot be satisfied as requested, and consequently a negotiation process with the client for the closest possible delivery time is proposed (1080), see Fig. 8. If this new deadline is accepted (old delivery time—red —is invalidated and new delivery time—green—is accepted) all demands are scheduled and executed in two separate commands as depicted in Fig. 9.

As a conclusion, the paper proposes an intelligent control solution for the production of radiopharmaceuticals composed of a hybrid control architecture together

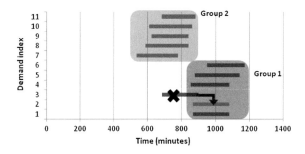

Fig. 8 Demand grouping and delivery time renegotiation

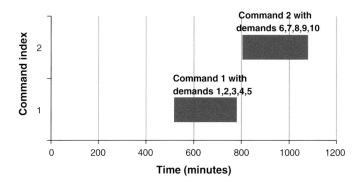

Fig. 9 Command execution after optimization and delivery time renegotiation

with a demand optimization sequence which groups demands into commands and orders.

Future research will cover the directions: (i) testing the optimization sequence for a larger set of products and production horizon, (ii) analyse how computed schedule differs from actual execution, (iii) minimizing material loss and (iv) consider energy costs as an objective function for the optimization sequence.

Acknowledgment This work is supported by the Sectorial Operational Programme Human Resources Development (SOP HRD), financed from the European Social Fund and the Romanian Government under the contract number POSDRU/159/1.5/S/137390/.

References

1. Iverson, Ch. et al. (eds.): 15.9.2 radiopharmaceuticals. In: AMA Manual of Style (10th edn). Oxford University Press, Oxford, Oxfordshire (2007). ISBN 978-0-19-517633-9
2. Schwochau, K.: Technetium. Wiley-VCH (2000). ISBN 3-527-29496-1

3. Ell, P., Gambhir, S.: Nuclear Medicine in Clinical Diagnosis and Treatment. Churchill Livingstone (2004). ISBN 978-0-443-07312-0
4. http://www.sciencedaily.com/releases/2010/07/100708111326.htm
5. http://www.ansto.gov.au/_data/assets/pdf_file/0019/32941/Nuclear_Medicine_Brochure_ May08.pdf
6. Mas, J.C.: A patient's guide to nuclear medicine procedures: English–Spanish. Soc. Nucl. Med (2008). ISBN 978-0-9726478-9-2
7. Medema, J., Luurtsema, G., Keizer, H., Tjilkema, S., Elsinga, P.H., Franssen, E.J.F., Paans, A. M.J., Vaalburg, W.: Fully automated and unattended [^{18}F] fluoride and [^{18}F] FDG production using PLC controlled systems. In: Proceedings of the 31st European Cyclotron Progress Meeting, Zurich (1997)
8. Kusiak, A.: Intelligent Manufacturing Systems. Prentice Hall, Englewood Cliffs (1990). ISBN 0-13-468364-1
9. Tsai, W.T.: Service-oriented system engineering: a new paradigm. In: Proceedings of the 2005 IEEE International Workshop on Service-Oriented System Engineering (SOSE'05). IEEE Computer Society (2005). 0-7695-2438-9/05
10. De Deugd, S., Carroll, R., Kelly, K.E., Millett, B., Ricker, J.: SODA: service-oriented device architecture. IEEE Pervasive Comput. 5(3), 94–96 (2006)
11. McFarlane, D., Giannikas, V., Wong, C.Y., Harrison, M.: Product intelligence in industrial control: theory and practice. Ann. Rev. Control 37, 69–88 (2013)
12. Van Brussel, H., Wyns, J., Valckenaers, P., Bongaerts, L., Peeters, P.: Reference architecture for holonic manufacturing systems: PROSA. Comput. Ind. (Special Issue on Intelligent Manufacturing Systems) 37(3), 255–276 (1998)
13. Leitao, P., Restivo, F.: ADACOR: a holonic architecture for agile and adaptive manufacturing control. Comput. Ind. 57(2), 121–130 (2006)
14. Raileanu, S., Anton, F., Iatan, A., Borangiu, Th., Anton, S., Morariu, O.: Resource scheduling based on energy consumption for sustainable manufacturing. J. Intell. Manuf. Springer (2015). Print ISSN: 0956-5515, On line ISSN: 1572-8145, doi:10.1007/s10845-015-1142-5
15. Novas, J.M., Bahtiar, R., Van Belle, J., Valckenaers, P.: An approach for the integration of a scheduling system and a multiagent manufacturing execution system. towards a collaborative framework. In: Proceedings of the 14th IFAC Symposium INCOM'12, Bucharest, pp. 728–733, IFAC Papers OnLine (2012)
16. www.constraintsolving.com/solvers. Consulted in Oct 2015
17. ILOG: (2009). www-01.ibm.com/software/websphere/ilog/. Consulted in Oct 2015

Improving the Delivery of a Building

Vince Thomson and Xiaoqi Zhang

Abstract There is a great difference in the effectiveness of product industries and the construction industry in the use of technology for better execution of development projects. This paper discusses some recent technologies that are able to create a better environment where construction companies can partner to share the cost-benefit of product information in order to optimize the cost and timeliness for building construction. The basis of this improvement is the use of active products that can update their own information. Products as intelligent operators have the greatest effect by automatically changing product data.

Keywords Construction · Part tracking · Intelligent object · Change management

1 Developing New Products

1.1 Product Industry

In today's environment, a product manufacturer usually designs and makes a complete product, or at least, controls all the processes for design and manufacture. An integrated product development team (IPDT) creates a design in a CAD (Computer Aided Design) system and manages the design with a PDMS (Product Data Management System) and/or a PLM (Product Lifecycle Management) system. This allows product management from the point of view of product design, part lists, part creation processes and assembly. Product information is integrated with ERP (Enterprise Resource Planning) systems (also known as Enterprise Resource Management Systems (ERMS)) to control product manufacture and supply chain processes. The cost and timeliness of product creation are predicted and controlled by the IPDT.

V. Thomson (✉) · X. Zhang
Mechanical Engineering, McGill University, Montreal, Canada
e-mail: vince.thomson@mcgill.ca

© Springer International Publishing Switzerland 2016   21
T. Borangiu et al. (eds.), *Service Orientation in Holonic and Multi-Agent Manufacturing*, Studies in Computational Intelligence 640,
DOI 10.1007/978-3-319-30337-6_2

It is the integration and sharing of information that allow the control of cost and timeliness [3]. A company that designs and makes its own products controls the cost-benefit of creating and using product information and of controlling the schedule and supply chain to make timely products. An example is Apple, who completely controls the designs of processor, hardware, operating system, application software and related cloud services as well as retailing stores [16]. For products that are developed in partnership, the OEM (Original Equipment Manufacturer) creates a team along with terms and conditions that incentivize the IPDT to share information and to optimize cost-benefit across product components.

In today's product information systems, products behave as both passive and active actors regarding their information [15]. Usually, a product does not carry its own information, but it carries identification (bar code, RFID tag (Radio Frequency IDentification), etc.), which provides a connection to its information stored in a set of databases. Much of this information is 'active' in that a change to a product datum automatically triggers changes to other data about the product in terms of design, manufacture or supply, assembly process, maintenance, etc. [5]. This level of integration of product data and associated automation with regard to product change has been continuously developed over the past 50 years since the first CAD systems. Nevertheless, although integration of product information is beneficial to the design process, most of the benefits accrue during the later stages of a product life cycle: part production, product assembly and maintenance, that pay for the high cost of integrating product information.

1.2   Construction Industry

Even though CAD systems for architecture, engineering and construction (AEC) were developed about the same time as CAD systems for other types of products, the drive for integration of product information and the accompanying automation for managing change have not occurred in the construction industry as it has in product industries [10]. The reason for this is that most of the time the overall process: the design, part production and building assembly procedures, are not controlled by any one company or group of companies, and thus, there is a mismatch between the cost of the integration of product information during the design phase and the benefits which occur mainly downstream during construction and maintenance. Typically, you have the architect and engineer who develop the design, and the general contractor, major contractor, and sub-contractor, who actually cooperate to construct a building, but in the process all partners act as individuals who optimize their own cost and activities due to marginal incentives to optimize overall cost and respect timeliness.

As a consequence, there is rarely an overall product model, and at best, there may be small, limited models created by individual companies. In general, the construction industry does not use IPDTs, rarely has group incentives to drive project goals, and does not use arrangements to collectively achieve goals for building construction. One major reason is due to different operating environments for product and construction industries. The construction industry uses contract bidding at each tier of the construction process, which actually inhibits cooperation towards integrating information.

There are some exceptions. One such example is the construction of the Sutter Medical Center (136 bed) project in Castro Valley, California. Project management was able to group together partners such that they shared project profits. This allowed the formation of an IPDT that created a single building information model (BIM), which facilitated the evaluation of alternative designs as well as the management of design and schedule change [12]. As a consequence, building permit approvals were obtained in record time; building construction was delivered at the guaranteed minimum price, and had an accelerated schedule that was 30 % faster than a conventional schedule [6]. Construction change orders and requests for information for structures were less than 15 % of estimates for comparable hospital projects in California [2].

1.3 Paper Objectives

The goal of this paper is to convey that some recent technologies are able to create a better environment in the construction industry where companies can partner to share the cost-benefit of product information in order to optimize the cost and timeliness of building construction, and that the basis of these improvements is the use of active products that can update their own information. Section 2 describes technologies that have recently entered the market or that their price and functionality have been significantly improved. Section 3 discusses the advantage of the technologies with regard to reducing project cost and improving timeliness. Conclusions are stated in Sect. 4.

2 Technologies for Improving the Delivery of a Building

2.1 Product Information System

The heart of any construction activity is the bill of materials for a building from which a schedule for part creation or supply and building assembly is made. ERP systems store information about a building and the pieces and devices in it. ERP systems also accommodate change management in terms of part alternatives and modifications to schedules and suppliers. The bill of materials flows from a building design. There are many systems that can design a building and create a BIM. The BIM is a digital representation of the physical and functional characteristics of a facility. The goal is to use the BIM as a shared knowledge resource for information about a facility forming a reliable basis for decisions during its lifecycle (conception to demolition) [7]. Moving from concept to BIM for a large building requires the cooperation of many partners and the integration of much data from each partner.

The greatest issue with regard to managing product information stored in a BIM, ERP, PDMS, etc. is defining, confirming and executing change. Recent research into change management has shown that it is possible to define the chain of change propagation and to manage the execution of changes in a design and manufacturing plan [17]. The management of the change process by the IPDT is another issue, and not covered here. Systems that can manage procedures for changing a design greatly reduce the effort for all the partners; for, one of the major cost drivers during product development is boundary management, i.e., the use of coordination mechanisms to assure the delivery of material and information across organizational boundaries: internal and external [3]. The process of managing change uses a lot of resources and causes much delay, where there can be instances of a design document being exchanged over ten times until it is finalized [14]. The more automation can be brought to bear on this process, the better. Thus, having available a methodology modifying product data when there is a change to the product or its circumstances greatly reduces the cost of the production process.

ERP systems are readily available. They range from generic, highly functional systems, such as SAP, to AEC specific systems, such as COINS, which provides traditional ERP functionality along with design functions to create a BIM and applications for construction specific analysis. The price for such systems is continuously being reduced. Even though this is resulting in wider adoption, the construction industry still does not have the adoption rate of product industries [1]. Thus, it is about taking advantage of automation to reduce one of the main cost drivers in building construction: change management.

2.2   Part Identification and Tracking

Delivering the thousands of parts for a large building is a very difficult task. This usually involves hundreds of partners and thousands of deliveries into a cramped construction site. RFID is a great tool that allows the identification of parts and their tracking at the manufacturer and at the construction site. There are several options.

- There is simple identification of parts with an RFID reader when using an RFID tag. The tag can be found at any time; however, the RFID reader only determines that the part is in its range. It does not provide location. A short search is needed.
- An RFID tag can have memory where the user can store not only part identification, but also other design and construction parameters.
- The majority of RFID tags are passive, i.e., the energy to transmit data comes from the radio frequency signal scanning a tag, and so, range is limited. There

are active (scheduled transmission) and passive (on demand) battery-assisted tags where local power allows transmission of radio signals over greater distances. Battery-assisted tags are more expensive, but allow automated tracking of parts when there is a lot of movement and/or distances are large.

Several companies use RFID tracking to their advantage. Examples are given below.

Armtec is a company that supplies precast, corrugated steel and HDPE products. In 2010, Armtec was awarded a $Cdn 43-million contract from the Toronto Transit Commission (TTC) to supply 58,000 subway tunnel, liner segments over a two year period. "This job requires handling hundreds of pieces per day," said Phil Sheldon, operations manager. "The tracking of those TTC pieces would have been very difficult to do without RFID technology." [8] Armtec used a combination of barcode and RFID embedded tags along with a GPS (Global Positioning System) to track and locate pieces, and to store the information.

At Armtec's Woodstock site alone, the estimated cost of missing inventory was $Cdn 260,000 per year, including $Cdn 60,000 in annual penalties for late deliveries where suppliers were fined up to $Cdn 3000 per hour for causing delays at job sites that resulted in construction crews being idle [8]. In addition to reducing missing inventory and late delivery penalties, Armtec saved money by reducing the many hours spent searching for pieces in its 20-ha site. TTC is now using the technology to identify and locate damaged tunnel segments quickly, thereby saving time and providing better maintenance.

Similarly, JV Driver, who provides industrial construction services to the resource sector in western Canada, uses RFID tracking technology to find pieces on its construction sites, where piece average search times were reduced from 30 to 5 min [11].

During the building of the 200,000-m^2 Metlife Stadium, Sanska USA Building (Parsippany, New Jersey) used RFID to track 3200 precast pieces to form the 84,000-seat bowl of the stadium. A just-in-time delivery system where tracking information was fed into a BIM meant that Skanska could identify which pieces had been manufactured and their quality status, what jobsite areas needed to be prepared, and what pieces were already incorporated into the building [9]. Skanska did not need to establish a laydown yard, but instead relied on a small holding area. For the precast pieces, Skanska estimated the tracking solution accelerated the construction schedule by 10 days and created $US 1 million in savings [9].

2.3   Part Location

Objects can be identified and located by using real time locating systems (RTLS). These systems can use active or passive RFID or infrared tags along with multiple readers. Items can be identified and positions located as precise as 5 cm using algorithms such as triangulation and technologies such as ultra-wide band [4].

Absolute positions can be obtained by use of a GPS or local tags with known position.

A RTLS reduces the search time for parts, allows for 1 piece flow from supplier to assembler, and permits time information for placing a part into its final location. Continuous part location reduces the labour due to continuous checking of delivery and assembly schedules as well as the supply chain for parts. An RTLS can be used as part of an information change process to update product information.

In the $US 10-billion Clair Ridge project by BP, a global oil and gas company, hundreds of suppliers took part in the construction of a new offshore oil platform by delivering components from two consolidation centres in Europe to the building site in South Korea. BP used an RTLS system consisting of both RFID and GPS technologies to track parts and to minimize delays. The RTLS system allowed real time visibility of parts moving from suppliers to the construction site, and helped BP to reach zero material loss and to significantly improve the planning process [13].

2.4 Self-organizing Wireless Microrouters

Routers are devices that use IEEE 802.11 communication protocols to allow connections among devices and between devices and networks. Of interest are those devices that can form self-organized networks, i.e., automatically making their own network. They do not need to be connected to a formal network or the Internet. Each device acts as a wireless microrouter that can establish peer-to-peer connections and relay messages from one peer to another. So, peers form a network and they relay messages such that they reach their final destination. Thus, a microrouter only needs to be connected to one other router, not to all routers. Self-organizing wireless microrouters form a true network topology, not a star. If one device is connected to a network, all microrouters have access. If one device fails, the other devices repair the network.

At a construction site, devices with an RFID tag and self-organizing wireless microrouters along with access to an RTLS can form a self-identified network, i.e., a peer-to-peer network where each device knows the identity and location of every other device. One device can download a location map of all devices into a database on schedule or on command. Part locations can be checked against the construction plan (BIM) and schedule for any deviations, and the data in an ERP can be updated.

2.5 Self-integrating Objects

Traditional networks of devices are defined by standards for connectivity. The host system provides this connectivity, and the software for communication protocols, for access to data, and for execution of special algorithms is provided by applications in the host system. These applications are loaded into the host, and if modifications occur in an attached device, a new application needs to be acquired.

A new approach is to have devices that are intelligent, understand the topology required for network integration, able to communicate on a network, and contain their own applications. In this approach, devices deliver applications to the host system during integration. If a device is changed (options made operative, upgrade), the device acquires a new application itself or self-modifies its present application. Then, negotiation is made with the host to install the new application.

This architecture makes a system device (product) responsible for its own information in addition to its own integration into a system. It provides a product with the communication and negotiation capability to resolve issues with the host system.

Consider an HVAC system in a large building. There is usually a central system which receives information from local controllers throughout the building to balance the overall system and to create the desired environment. Each controller usually has multiple sensors and functions. Although the original building plan determines the number of sensors, area controllers and their functions, allocates the number of interface connections, and sets the specifications for the host software, there are always changes: more or different sensors, different functions for area controllers, as well as significant change to control software. This causes a sea of changes in the HVAC system, which adds considerable effort to the building construction and maintenance.

If each area controller had a self-contained application and could effect its own changes by negotiating with the central system, then the impact of change would be highly reduced due to the automated change mechanism. Even if different devices are used than planned, the automated negotiation and integration of interfaces greatly reduces the effort due to change.

3   Discussion

At the beginning of the paper, the difference between product industries and the construction industry with respect to the greater use of technology by product industries for product data integration in order to reduce overall cost and to improve the timely delivery of products was described. In the construction industry, the capability of part tracking to reduce material handling cost and to deliver better timeliness for building assembly was shown. However, it is the use of technology that allows products to change their own product data that greatly improves the use of product information in the construction industry, since this type of automation greatly reduces the amount of labour and time. With this technology, collaboration among partners increases since the improvement in construction information has increased benefits for all partners.

The updating of part location can provide the status of part delivery, jobsite location and final position in building assembly. Moreover, besides providing this data, parts need to act as intelligent operators and transparently change their data in an ERP or BIM. It is the automated initiation of change by a building piece or device through networked systems that reduces the cost of building assembly, maintenance and future data integration. Technologies such as self-identifying objects, RTLS, and wireless microrouters provide the basis for making device initiated change a reality. However, it is being a self-integrating object that allows a product to be an intelligent operator, which can respond to its environment and update its information.

When reviewing projects such as the construction of the Sutter Medical Center, there are three main reasons for achieving the high level of success: the agreement by partners to a single, collaborative contract which outlines partner responsibilities and the sharing of cost and benefits; the use of technologies like BIM; and the use of Lean project principles [2, 12]. This paper has discussed the use of technology to automate change in design and operations. It is also clear that this use of technology needs to be built upon best practices in the forming of partnerships and in project management.

4 Conclusion

There have been many successful projects that use an IPDT as well as create a BIM to share building information, and thus, they have been able to create synergies that have reduced cost and have delivered buildings on time. Unfortunately, to date, these cases have been limited and the methodologies have not been widely adopted. This paper has described some technologies that allow building pieces and devices to be intelligent operators such that they can act to automatically and transparently update product data in order to improve cost and timeliness during building construction. The technologies are inexpensive and readily available. Overall, the construction industry needs to combine the use of smart devices as intelligent operators and the use of best management practices during building projects in order to improve productivity.

References

1. Ahmed, S., Ahmad, I., Azhar, S., Mallikarjuna, S.: Implementation of enterprise resource planning (ERP) systems in the construction industry. In: ASCE Construction Research Congress—Wind of Change: Integration and Innovation, Honolulu, 1–8 Mar 2003
2. Aliaari, M., Najarian, E.: Sutter Health Eden Medical Center: structural engineer's active role in an IPD project with lean and BIM components. Struct. Mag., 32–34 (2013, Aug)
3. Anaconda, D., Cladwell, D.: Improving the performance of productivty teams. Res. Technol. Manag., 37–43 (2007, Sept–Oct)
4. Connell, C.: What's new in real-time location systems? Wirel. Des. Dev. 21, 36–37 (2013)

5. Jun, H.B., Shin, J.H., Kim, Y.S., Kiritsis, D., Xirouchakis, P.: A framework for RFID applications in product lifecycle management. Int. J. Comput. Integr. Manuf. **22**(7), 595–615 (2009)
6. Khemlani, L.: Sutter Medical Center Castro Valley: case study of an IPD project. AECbytes, 1–11, 06 Mar 2009. http://www.aecbytes.com/buildingthefuture/2009/Sut-ter_IPDCaseStudy.html. Accessed 27 Aug 15
7. National BIM Standard: Frequently asked questions about the National BIM Standard. National BIM Standard (United States). Nationalbimstandard.org. Accessed 24 Aug 2015
8. National Research Council Canada: GPS reduces construction cost. Dimensions (6) (2011)
9. Palmer, W.D.: Tracking precast pieces: technology that works. Concr. Construction (2011, Oct)
10. Shen, W., Hao, Q., Mak, H., Neelamkavil, J., Xie, H., Dickinson, J., Thomas, R., Pardasani, A., Xue, H.: Systems integration and collaboration in architecture, engineering, construction, and facilities management: a review. Adv. Eng. Inf. **24**(2), 196–207 (2010)
11. Soleimanifar, M., Beard, D., Sissons, P., Lu, M., Carnduff, M.: The autonomous real-time system for ubiquitous construction resource tracking. In: Proceedings of the 30th ISARC, Montreal, Canada (2013)
12. Staub-French, S., Forques, D., Iordanova, I., Kassalan, A., Abdulall, B., Samilski, M., Cavka, H., Nepal, M.: Building information modeling (BIM) 'best practices' project report. University of British Colombia (2011). http://bim-civil.sites.olt.ubc.ca/files/2014/06/BIMBestPractices-2011.pdf. Accessed 27 Aug 2015
13. Swedberg, C.: RFID, GPS bring visibility to construction of BP oil platform. RFID J. (2013, May 08). http://www.rfidjournal.com/articles/view?10659. Accessed 27 Aug 2015
14. Thomson, J., Thomson, V.: Using boundary management for more effective product development. Technol. Innov. Manag. Rev., 23–27 (2013, Oct)
15. Trentesaux, D., Thomas, A.: Product-driven control: concept, literature review and future trends. In: Borangiu, T., Thomas, A., Trentesaux, D. (eds.) Service Orientation in Holonic and Multi Agent Manufacturing and Robotics, vol. 472, pp. 135–150. Springer, Heidelberg (2013)
16. Wharton: How apple made 'vertical integration' hot again—too hot, maybe (2012). http://business.time.com/2012/03/16/how-apple-made-vertical-integration-hot-again-too-hot-maybe/. Accessed 1 Oct 2015
17. Wynn, D., Caldwell, H., Clarkson, J.: Predicting change propagation in complex design workflows. ASME J. Mech. Des. **136**(8) (2014)

Repair Services for Domestic Appliances

Rachel Cuthbert, Vaggelis Giannikas, Duncan McFarlane and Rengarajan Srinivasan

Abstract There has been a trend of increasing levels of Waste Electrical and Electronic Equipment over the last few decades as the possibility for accessing repair of appliances has declined. Reducing prices of appliances has also generated a culture where the disposal and replacement with new appliances is the quicker and cheaper option, compared with repair. A number of key areas have been identified as important in helping to increase the number of appliances which may feasibly be repaired in the future. Of these, two key areas encompass the automation of repair of appliances, and the information requirements in order to achieve this. Within this paper, a demonstrator will be described which provides a step towards illustrating the potential of product intelligence and semi-automated repair.

Keywords Repair · Domestic appliances · Automation · Information requirements

1 Introduction

The level of waste generated through the disposal of domestic appliances has increased significantly over the past decades. In many cases, the items which are disposed of are in working order or could be repaired with little work. This adds to the increasing concern around the large amount of energy expended on the

R. Cuthbert (✉) · V. Giannikas · D. McFarlane · R. Srinivasan
Institute for Manufacturing, University of Cambridge, Cambridge CB3 0FS, UK
e-mail: rc443@eng.cam.ac.uk

V. Giannikas
e-mail: v.giannikas@eng.cam.ac.uk

D. McFarlane
e-mail: dcm@eng.cam.ac.uk

R. Srinivasan
e-mail: rs538@eng.cam.ac.uk

© Springer International Publishing Switzerland 2016
T. Borangiu et al. (eds.), *Service Orientation in Holonic and Multi-Agent Manufacturing*, Studies in Computational Intelligence 640,
DOI 10.1007/978-3-319-30337-6_3

'recycling' of materials and the manufacture of new goods. What is worrying is the irresponsible nature of product design where items may have built in obsolescence such that their lifetime is not as long as perhaps it could be. The intention of this is to fuel repeat business. Unfortunately, product costs have been driven down so much that, while repair used to be the norm, consumers now opt to buy new and throw away failed or unfashionable items.

The background of the Distributed Information and Automation Laboratory research[1] looks at key areas of relevance, namely, information requirements, quality, availability, sensing (condition of equipment etc.), automation and product intelligence. The research into the area of repair started with a master's level student project looking at Design for Repair in 2014 and this has been followed by an investigative scoping study over the past 10 months. The aim of this work is to significantly increase the number of domestic equipment that can feasibly be repaired.

This paper is structured as follows. We begin with a background section discussing the issues of waste, design, repair and obsolescence of domestic appliances, and the consequences that this has on the level of disposal of appliances. We then present a research agenda for the topic of repair of domestic appliances and the major challenges associated with it. In Sect. 4, we then present an intelligent product based demonstrator which our research team is currently working to study the technical challenges associated with repair.

2 Background

Over the last 30 years, the replacement of failed domestic appliances with new has become a relatively inexpensive, quick and easy solution. Simultaneously, repair has become more costly, time-consuming and inaccessible. The ease with which appliances can be disposed has increased while the separation and sorting of waste has become more 'responsible' [14]. Around 25 % of disposed domestic appliances are reported to be in working order. Often simple repairable faults or fashion 'whims' lead to appliance disposal and replacement.

Throughout the supply chain significant revenue may be gained from repeat business and the continued disposal of domestic appliances despite the negative materials and energy consequences (Fig. 1). Financial incentives are such that companies design in obsolescence to ensure future demand, with the damaging consequence of fuelling a throwaway culture. However, in cases where electronic devices have become obsolete many which are clean and functional can be reused if identified and sorted out by experts [7].

The UK generated 200 million tonnes of total waste in 2012. Of this, WEEE (Waste Electrical and Electronic Equipment) accounts for around 2 million tonnes

---

[1]http://www.ifm.eng.cam.ac.uk/research/dial/.

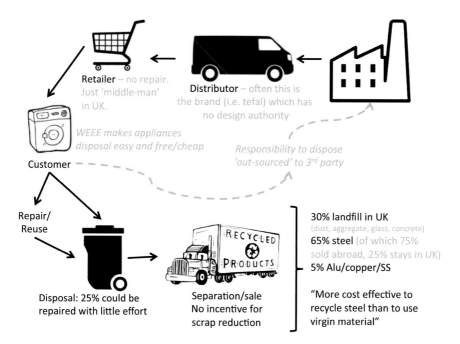

Retailer – no repair. Just 'middle-man' in UK.

Distributor – often this is the brand (i.e. tefal) which has no design authority

WEEE makes appliances disposal easy and free/cheap

Customer

Responsibility to dispose 'out-sourced' to 3ʳᵈ party

Repair/ Reuse

30% landfill in UK (dust, aggregate, glass, concrete)
65% steel (of which 75% sold abroad, 25% stays in UK)
5% Alu/copper/SS

Disposal: 25% could be repaired with little effort

Separation/sale No incentive for scrap reduction

"More cost effective to recycle steel than to use virgin material"

Fig. 1 Product supply chain and stakeholders

discarded by householders per year [4]. In real terms, estimates are such that for every tonne of consumer waste, 5 tonnes of manufacturing waste and 20 tonnes of resource extraction waste have also been generated [8].

A key question is whether all items need to go straight to the recycling stage rather than a greater number being reused or repaired. Repair, historically, was a far cheaper and more accessible option and electricity boards had shops which provided repair services as a means for them to sell electricity (through the purchase of their appliances). Furthermore, some repairs are regarded as being of inferior quality to remanufacturing options, with warrantees only covering the repaired component [8]. Unfortunately, the repair business has ceased to exist as electricity supply has become commonplace in residences, and as new appliances have become much more affordable.

Product design and replacement has a significant impact on WEEE generation and treatment. Where product replacement is fast, WEEE will increase dramatically in a short time, and then decrease rapidly creating peaks and troughs in waste treatment processing facilities [10]. One way of trying to reduce levels of WEEE due to product failure is by improving the design such that they are more repairable. Design for X is a key term that has been applied across many important areas. Ones which have significance to this work include:

- Design for Disassembly; the disassembly of parts, components and materials.
- Design for Recycling; recycling components, parts or materials.
- Design for Reuse; reuse of a product 'as is' or harvesting working parts or components for reuse, often in the form of repairs and replacements.
- Design for Remanufacturing; remanufacturing a product to be like new and then reselling the product, often in a different market [1].
- Design for Product life extension; requires tailored approaches for different product types [2].

However, this work seeks to research the possibility of Design for Repair in which the product is designed to be more easily repairable, and ownership is maintained throughout the repair process. Another way to decrease the number of appliances that are discarded is to improve the accessibility of repair.

3   A Research Agenda

3.1   Key Areas on Which to Focus the Research

We identify five areas that future research should deal with in the area of repair of domestic appliances:

1. *Economic requirements and business/contract models* to determine how repair, sales, and product use may be combined in such way that repair is managed and achieved more often in an after sales capacity.
2. *Design guidelines and material considerations*, which render products easier to diagnose faults, disassemble, repair and reassemble, potentially by the end-user.
3. *An information model to enable/support/enhance the repair/replacement/upgrade process*. This would ensure that the right information is available to the right player within the supply chain, such that repair may be achieved.
4. *Automation of repairs* in order to make repair a quicker, cheaper, more repeatable and accurate possibility.
5. *Standards and legislation*, which are key areas to support the above 4 areas.

Of particular interest in this context are the areas of automation and information:

- *Automation:* Design and development of basic automated repair functions. Automation could be of a collaborative nature where a person provides key intelligence and decision making, while a collaborative robot carries out key disassembly, test, and repair activities.
- *Information Management*: What information is required to enable/support/ enhance the repair/replacement/upgrade process for appliances. How does this differ if the repair is carried out by the consumer or by a robot? Can information from the appliance trigger the repair appointment, spare parts ordering, etc.?

Table 1 Factors related to the 'repairability' of appliances

Diagnosis	Disassembly	Repair	Reassembly
Symptoms	Type of operation	Identification of fault	Scope to re-assemble
Visual	Tool required Modularity	Is a part faulty? Identification of faulty part	Availability of parts and tools to complete this?
Mechanical	Ease of access	Can part be repaired?	What goes where?
Electrical	Is access not destructive?	Can the part be replaced?	Accessibility of repair site
Functional	Skill required for each operation	Where to order from if new part required?	Requirement for fixtures/jigs
Testing	Re-usability of parts/fixing	Skills to replace technically need a qualified engineer	Potential of error in re-assembly
Web based help	No of steps from each operation	Ease of replacement Cost of part	Safety of re-assembled unit
Manuals	Safety of operation	Tools to replace	
Sensor data	Accessibility	Access to replace	

3.2   Challenges and Shortcomings

It is not anticipated that one would attempt to adopt automation for all domestic appliance repairs in the first instance. The key question around which appliances should be repairable will depend on a number of factors, but clearly value will be one of them. The question then becomes how much we can lower the threshold of what is economically viable to repair, considering the very varied nature of domestic appliances from high to low value.

Further factors come into play when considering the 'repairability' of appliances.

These relate to how straightforward it is to diagnose the problem in the first instance, and then on a more practical level, how feasible it is to disassemble, repair and re-assemble the appliance in a safe and effective way. A number of factors considered in this area are shown in Table 1.

4   The Role of Product Intelligence

Product intelligence is a paradigm that could support both the information management and the automation challenges of repair. In this section, we review similar work in the literature and we present a demonstrator based on the product intelligence approach.

4.1 Relevant Work

Intelligent products, along with other similar paradigms, are argued to offer special benefits in middle-of-life services like maintenance and repair [9, 13, 15]. These benefits refer to the collection and gathering of item-based information about a product's use (using sensors embedded on the product itself), which can then be distributed to third parties and/or be used to detect abnormalities and failures.

This has led researchers to focus their existing work around remote diagnostics services enabled by intelligent products, which can be used to improve problem diagnosis, to improve condition-based maintenance and to schedule service personnel. In a domestic environment, there are examples of intelligent product developments for video game consoles [16], refrigerators [6] and washing machines [11]. It is also argued that product intelligence can lead to the development of smart appliances that can be used to improve the energy efficiency of modern houses [5]. Apart from houses, it has been shown that intelligent products can facilitate better repair and maintenance services for vehicles [5] and aircraft [3].

However, it is acknowledged that in the context of domestic appliances, there will be a cost/intelligence trade-off and the question here is really around what types of appliance make sense to be repaired, and what appliances lend them to benefiting from intelligence (Fig. 2).

Figure 2 shows two extremes of appliance characteristics. At the high value end appliances are typically one-off, the 'brand' is the engineering authority, they are hand-made, expensive, repairable and have a long lifetime. At the low value end they are produced in high-volume, the 'brand' is added to pre-built mass-produced items, their manufacture is automated, they are low-cost, non-repairable and have a short lifetime. What is key within this work is to determine how far from the left to the right the sweet spot may be pushed, or in other words, the reduction in value of appliances for which it is economically viable to effect their repair.

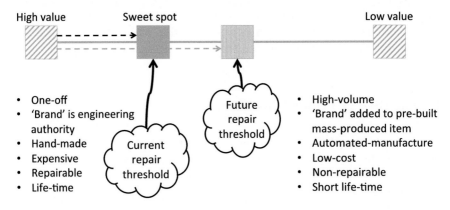

Fig. 2 Product supply chain and stakeholders

4.2 Repairing Appliances with Intelligent Products: A Demonstrator

In this section, we present a simple demonstrator our research team is currently working on to demonstrate some of the ideas behind repair of domestic appliances. These areas focus on the information requirements for product condition, usage, etc., and the automation of domestic appliances. Figure 3 illustrates the concept of the demonstrator.

There are two parts in this development. Firstly, the diagnostics part, where the problem of the appliance is diagnosed. The appliance, in our case a vacuum cleaner, equipped with sensors and memory collects information during its middle-of-life stage. This information is periodically checked against similar data from other appliances and troubleshooting processes in order to detect potential failures and their cause. A key question here relates to what information is required for the repair decisions to be made. The types of information required will be similar to that required in the provision of service contracts for complex engineering equipment [12]. Once a problem is diagnosed, the appliance makes a decision about its repair. Besides the numerous examples that discuss the details of the technical implementation of similar demonstrators (see for example [9]), we note here that, affordable and small equipment like a Raspberry Pi attached to the appliance (or even the appliance's own electronics) could be used for our purposes.

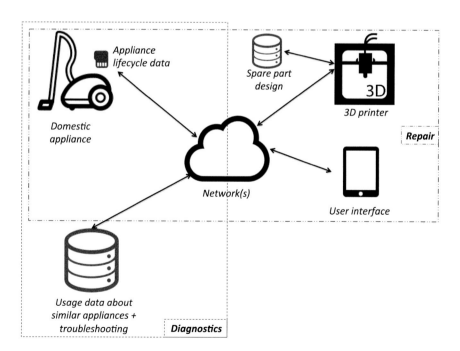

Fig. 3 Demonstrator

The second part of the demonstrator refers to the repair. A user, via a network (e.g. Internet, Bluetooth) can communicate with the appliance to receive instructions guiding them through the repair process. User-friendly interfaces could be designed for computers or even tablets and phones using apps. In certain cases, repair could be a simple process like the replacement of a filter. In more complex scenarios, an old spare part might need to be replaced with a new one.

Table 1 indicates a number of the issues which need to be considered in the process of diagnosis, disassembly, repair and reassembly of appliances. With the emergence of 3D printing, spare parts could be printed at low cost in designated locations or even inside the owner's house. Using this interface, a user would send a command for printing to a 3D printer, which will then find and use the spare part design over the Internet or in an allocated database. In this way, a customer could easily replace faulty spare parts and repair their domestic appliances faster and potentially cheaper.

5 Conclusions

Within this paper we have introduced the issues around the levels of waste generation, in particular, in the context of WEEE and domestic appliances. Built-in obsolescence, product cost, inaccessibility of repair have all led to a throwaway culture where replacement of products with new ones is seen as the quickest, cheapest and most reliable option. Key research areas, which seek to reduce the number of appliances that may feasibly be repaired, have been presented. Of these key areas, two main areas (information requirements and automation of repair) are being incorporated into a demonstrator which is described and which aims to illustrate the possibility of increased repair of appliances through more easily accessible product information, diagnosis and repair possibilities.

References

1. Arnette, A.N., Brewer, B.L., Choal, T.: Design for sustainability (DFS): the intersection of supply chain and environment. J. Cleaner Prod. **83**, 374–390 (2014)
2. Bakker, C., Wang, F., Huisman, J., den Hollander, M.: Products that go round: exploring product life extension through design. J. Cleaner Prod. **69**, 10–16 (2014)
3. Brintrup, A., McFarlane, D., Ranasinghe, D., Sanchez Lopez, T., Owens, K.: Will intelligent assets take off? Toward self-serving aircraft. IEEE Intell Syst **26**(3), 66–75 (2011)
4. Department for Environment, Food & Rural Affairs UK statistics in waste, March, Available online at https://www.gov.uk/government/statistics/uk-waste-data (2015)
5. Främling, K., Holmstrom, J., Loukkola, J., Nyman, J., Kaustell, A.: Sustainable PLM through intelligent products. Eng. Appl. Artif. Intell. **26**(2), 789–799 (2013)
6. Främling, K., Loukkola, J., Nyman, J. and Kaustell, A.: Intelligent products in real-life applications. In: International conference on industrial engineering and systems management. Metz, France, May (2011)

7. Kang, H.-Y., Schoenung, J.M.: Electronic waste recycling: a review of U.S. infrastructure and technology options. Resour. Conserv. Recycl. **45**, 368–400 (2005)

8. King, A., Burgess, S., Ijomah, W., McMahon, C.: Reducing Waste: repair, recondition, remanufacture or recycle? Sustain. Dev. **14**(4), 257–267 (2005)

9. Kiritsis, D.: Closed-loop PLM for intelligent products in the era of the internet of things. Comput-Aided Des **43**(5), 479–501 (2011)

10. Lu, B., Liu, J., Yang, J., Li, B.: The environmental impact of technology innovation on WEEE management by Multi-Life Cycle Assessment. J. Clean. Prod. **89**, 148–158 (2015)

11. Lopez, T., Ranasinghe, D., Patkai, B., McFarlane, D.: Taxonomy, technology and applications of smart objects. Inf. Syst. Front. **13**, 281–300 (2011)

12. McFarlane, D., Cuthbert, R.: Modelling information requirements in complex engineering services. Comput. Ind. **63**, 349–360 (2012)

13. Sallez, Y., Berger, T., Deneux, D., Trentesaux, D.: The lifecycle of active and intelligent products: the augmentation concept. Int. J. Comput. Integr. Manuf. **23**(10), 905–924 (2010)

14. WRAP.: Realising the reuse value of household WEEE (2011)

15. Wuest, T., Hribernik, K., Thoben, K.D.: Accessing servitisation potential of PLM data by applying the product avatar concept. Production Planning & Control (2015), forthcoming (2015)

16. Yang, X., Moore, P., Chong, S.K.: Intelligent products: from lifecycle data acquisition to enabling product-related services. Comput. Ind. **60**(3), 184–194 (2009)

End-of-Life Information Sharing for a Circular Economy: Existing Literature and Research Opportunities

William Derigent and André Thomas

Abstract Intelligent products carrying their own information are more and more present nowadays. A lot of research works focuses on the usage of such products during the manufacturing or delivery phases. This led to important contributions concerning product data management in the framework of HMS (*Holonic Manufacturing Systems*). This paper aim is to: (1) make a review of the major contributions made for EOL information management (data models, communications protocols, materials, …) in the framework of a circular economy, (2) have a first overview on the industrial reasons explaining why these systems are not widely implemented. This previous points help to highlight potential research directions to develop in the near future.

Keywords Circular product · EOL data management · Closed-loop supply chain

1 Introduction

Our current economy is based on a linear model "*extract-make-consume-throw*" where natural resources and energy are used to produce goods that become most of the time non-utilized waste. This business model, based on unlimited natural resources and on a strong capacity to recycle wastes, rapidly becomes obsolete in a context of finite resources and massively uncollected wastes and thus causes important economical, social and ecological problems. The growth of world population will inevitably lead to an increase of the demand in base materials and thus induce a shortage of natural resources and an increase in costs. The report *The limit to growth* written by MIT members invited by the Club of Rome considers in 1972

W. Derigent (✉) · A. Thomas
CRAN-CNRS UMR 7039, Campus Sciences, BP 70239,
54506 Vandoeuvre-lès-Nancy Cedex, France
e-mail: william.derigent@univ-lorraine.fr

A. Thomas
e-mail: andre.thomas@univ-lorraine.fr

© Springer International Publishing Switzerland 2016
T. Borangiu et al. (eds.), *Service Orientation in Holonic and Multi-Agent Manufacturing*, Studies in Computational Intelligence 640,
DOI 10.1007/978-3-319-30337-6_4

41

that the earth growth limits will be reached in the next hundred years if this current model is maintained.

An alternative to this linear model is to promote a model defined as *"circular"*, mainly a system where products would be designed to be used a long time and be easily recycled at their end of life. In this model, the product may have several different usage phases with different missions. To maintain a good performance level, the product could be updated between these different phases. If the cost is too high, it will be dismantled and its components would be reused to equip other products at a lower price. When components become obsolete, they are recycled to new raw materials that will be used to product new goods. This model, particularly different from the current one, constitutes nonetheless a credible alternative because based on a transposition of the very efficient natural model.

Companies willing to adopt a circular model usually follow a 5-step process [1]:

1. **Develop new business models**: in a circular model, the sales revenue of a company is not anymore linked to the quantities of sold products, but to services provided to customers. New development opportunities will appear on the different reprocessing loops;
2. **Develop new partnerships**: The circular model is based on the hypothesis that wastes from a company can become materials for another one. Setting up new partnerships can develop industrial symbiosis, such as the industrial site of Kalundborg, Danemark [1];
3. **Design and set up a closed-loop supply chain**: Recycling loops must be supported by associated logistics. Usually, a supply chain conveys products from manufacturer to customer. In a circular economy, companies have to manage the dismantlement process along with the diverse materials and product flows in a closed cycle. As a result, setting up a circular economy is equivalent to adding a loop to a classic supply chain. The key difference between these models is the End-Of-Life (EOL) management of products, which then becomes crucial for an efficient circular model. The activity of managing the products in their EOL is often referred to as *reverse logistics* (see Fig. 1). The combination of forward and reverse logistics results in a closed-loop supply chain [2].
4. **Design "circular products"**: products moving in a circular economy must be designed to ease their maintenance, remanufacturing or recycling processes. At the same time, companies adopting a service economy seek to increase their products' useful life by making them more reliable. *Eco-design* refers to the process aiming to develop a product integrating constraints issued from the circular economy.
5. **Manage company performance**: A company adopting a circular model needs to define new performance indicators to drive its performance, classical ones being not sufficient. Proposed indicators are either linked to product performances or supply chain performances. For example, the ISO 22628 standard

---

[1]for more information, please refer to http://www.ellenmacarthurfoundation.org/fr/case_studies/la-symbiose-industrielle-de-kalundborg.

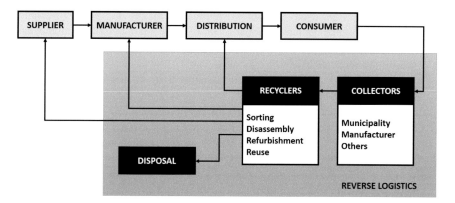

Fig. 1 The closed-loop supply chain (from [3])

defines indicators like the ***recyclability rate***, which is the part of product which can be reused or recycled. In addition, the ***used product processing time*** from its collect to its transformation is another relevant indicator that characterises the closed-loop supply chain performance.

As a result, to set up an efficient circular economy, EOL management is a crucial step because decisions taken during this one have a maximum impact on the recyclability rate of products. The performance of a closed-loop supply chain is essentially driven by the following factors [1]: (1) Products are not designed to be "circular", (2) Quantity and quality of used products are variable and not predictable, (3) Product information on used products is not sufficient. While problem *1* is more related to product design, problem *2* is clearly related to the closed-loop supply chain domain [2] and problem *3* is linked to information technology. This paper considers this last problem and aims to provide a clear overview of the actual research works in this area. Section 2 contains a short review on the solutions proposed by the intelligent product community to solve the lack of information during the EOL management phase. Then, Sect. 3 is dedicated to potential reasons explaining the non-emergence of the "EOL product holon".

2 Review on Solutions for Efficient EOL Data Management

End-Of-Life management involves options available to product after its useful life. [1, 4] illustrate five product recovery options:

- **Repair and reuse**: the purpose is to repair the product and to return it in working order. The quality of the repaired products could be less than that of the new products;

- **Refurbishing**: the used product is renovated thanks to the disassembly, inspection and replacement of some of its parts or sub-assemblies (also called *modules*). The quality of the refurbished product is brought to a specified level fixed by the recycler [5];
- **Remanufacturing**: the product is disassembled up to the component level, which are all inspected and replaced, in order to bring the used product up to quality standards as rigorous as those for new products;
- **Cannibalization**: Only a relatively small number of reusable modules from the product are preserved, to be used in any of the three operations mentioned above;
- **Recycling**: it is the last option, done when none of the above is possible. It consists in reusing materials from used products in the production of new goods.

In addition, parts and materials that could not be recovered by any of the above five operations will be **disposed** in accordance with safety and environmental regulations.

References [6–10] argue that intelligent products would help to ensure an information continuum from the BOL (Beginning of Life) phase to the EOL one, and consequently, more information would be available at the product end of life, and would help to optimize the EOL process. In the framework of the European project PROMISE, [3] emitted a list requirements that must be fulfilled by an EOL information system:

- *Req. 1*: Ability to uniquely identify and track products at the item level throughout the supply chain;
- *Req. 2*: Ability to provide relevant identity information associated with the product;
- *Req. 3*: Ability to provide information to determine the "current state" of the product;
- *Req. 4*: Ability to update product information as it changes throughout the product's lifecycle;
- *Req. 5*: Ability to provide instructions that enable the system to automatically route products through the product recovery processes;
- *Req. 6*: Ability to provide decision support at various stages of the product recovery processes.

Moreover, information conveyed with a product item enduring a EOL process should not disappear with the initial product item, but should be inherited by its derivative products (sub-assemblies, parts and raw materials obtained from it). As a result, a seventh requirement is proposed to complete the previous list:

- *Req. 7*: Ability to bequeath product information to its derivative products.

Table 1 proposes a short review on existing major approaches with intent to identify potential research lacks. For each of the previous requirements, a study has been made on 4 different aspects:

Table 1 Review of existing research works and materials for efficient product information retrieval (*gray = not fulfilling completely the requirement, hatch = not relevant regarding the requirement*)

Req.	Data	Communication	Decision	Materials
Req.1	EPC Code [13], ID@URI [14], Bio-identification [15], chemical identification [16]	DIALOG [17], WWAI [18], EPCIS [19]		Barcode, RFID (Pass./Act.), Wireless Sensor Networks (WSN), communicating materials [9]
Req.2				
Req.3	O-MI [20],O-DF [21] SOM [22], PRONTO [23], PLCS [24]			Active RFID, WSN, communicating materials
Req.4		Synchronisation techniques [25, 9, 10]		Barcode, RFID (Pass./Act.), Wireless Sensor Networks (WSN), communicating materials
Req.5	MIMOSA [26]	DIALOG, WWAI, EPCIS	None	
Req.6			No HMS-related approach [2]	
Req.7	Aggregation [27]	Data dissemination [28, 29]		WSN, communicating materials

- **Data modelling**: EOL processes need data (mainly product data). This aspect deals with data modelling, and lists all related research works trying either to identify or formalize data required by EOL processes;
- **Communication architecture**: In EOL, communication is an important feature to take into account, because EOL processes should be able to retrieve information located on the product and/or external databases. This aspect lists all related research works on communication architecture, data synchronisation and aggregation;
- **Decision making**: a good EOL management must be based on efficient decision-making processes, capable of processing all EOL data and choosing the best recycling alternative. This aspect deals with all works around decision-making for EOL;
- **Materials**: HMS are based on product holons, composed of an informational coupled with a physical part [11, 12]. This aspect deals with product technologies (industrial solutions or prototypes) that could support the concerned requirement.

Conclusions drawn from this short survey are interesting: first, during past years, numerous research has been made on product lifecycle data management, and many could be applied on EOL. Almost all requirements are mostly achieved, meaning

Product Lifecycle Information Management Systems (PLIMS) did attain a important level of maturity. Nevertheless, some cells of the table (colored in gray) are still to investigate:

- *Req. 4 "update product information as it changes"—Communication*: in this area, synchronisation techniques are important to ensure that the informational part is always a good representation of the physical one. The work of [10] exploits the O-DF/O-MI standards to present one solution of synchronization. Until now, it is the only work trying to achieve this aim, and does not provide any test case or deep study;
- *Req. 5 "provide instructions"—Decision*: providing the right instructions to EOL processes may suppose that the product is capable of taking decisions. As far as we know, applying a product-driven vision (which is a branch of HMS) to the EOL management has not been performed until now.
- *Req. 6 "provide decision support at various stages"—Decision*: as illustrated in [30], numerous methods & metrics exist to evaluate the different EOL alternatives. However, none of these methods is based on HMS approaches, or even MAS ones. The application of the HMS control paradigms on EOL management has not been extensively studied and an important application field is still unexplored;
- *Req. 7 "Bequeath the information to derivative products"—Data & Communication*: when a product is dismantled, and its derivatives reused, information should not be lost. It should be transformed, recombined or divided. [27] address the problem of product data aggregation when combining several products together. References [28, 29] design a data dissemination process applied on communicating materials, to ensure no data is lost after a physical transformation of the material. Making data robust to physical transformations, capable of automatic aggregation or dissemination is a key to ensure a continuous information flow all over the EOL processes, and it has not been investigated widely until now.

If some research questions are still to be studied, technologies and methods are already available to set up efficient EOL data management systems. However, currently, EOL management is still mainly recycling (which is the most energy-consuming EOL alternative, and thus the less cost-effective) and product information is most of the time non fully exploited, because unavailable [4]. In the next section, some reasons potentially explaining this report are discussed.

3   Discussions on the Problematic Non-emergence of EOL Product Holon

The previous review showed that currently, efficient technologies and methods are available to set up primary EOL data management systems. Some are old, seem to be efficient but they nevertheless did not push forward the development of EOL data management solutions. One may then wonder why the EOL product holon does not naturally emerge, pushed by industrial needs.

3.1 Development of the Needed Infrastructures

Managing data all along the product lifecycle is a very hard task: numerous stakeholders, different locations, different standards, long lifecycles ... make circular data management difficult to achieve. Moreover, along a given reverse supply chain, the reprocessing decision is highly distributed and depends on the technical, social, economical or even geographical product context (available resources, allocated maximum reprocessing costs, legislation, ...). Developing such infrastructures is equivalent to build collaboration platforms between companies associated in a same reverse supply chain, at the product level, in which a given item would send information about its evolution, and seek information about its environment in order to take EOL decisions. This is clearly an important challenge, and the development of cloud solutions, of wireless networks should ease the development and deployment of the product monitoring infrastructure. There will always remain some interoperability or data accessibility problems, but in that case, products could be equipped with memory-on-board to store data, that would be uploaded in the cloud when connection would be restored.

3.2 No Adapted Business Models

Building these new infrastructures might require major investments that will be driven by new business models. However, most of our modern companies are not fully concerned by sustainability problems and do not consider that switching to circular economy might bring benefits. As a result, the development of closed-loop supply chains is slow: companies do not want to invest in networks that won't provide money. Nevertheless [1] do illustrate some interesting examples of companies who adopted a circular model and made important benefits by recycling their products or moving towards service economy. These new business models will inevitably force companies to develop EOL management strategies required to extend their products' useful life, and maximise their benefits. The shift from our current economy to the circular economy may be faster with appropriate proof of return on investment.

3.3 No Proof of Possible Return on Investment

ROI (Return On Investment) is a key to persuade investors to adopt the circular economy. Some research works provide some evidences that efficient EOL management can bring benefits in term of bullwhip effect, inventory variance or product cost [31, 32]. Moreover, companies will require deployment methodologies and tools to evaluate qualitatively and quantitatively the benefits of such systems. Obviously, ROI will not be the same for every actors of the closed-loop supply chain, and these new tools will also help companies to identify which portions of the chain are the most or the least profitable. In certain cases, some links of the closed-loop supply chain might be unprofitable but absolutely required for the chain. Some business dispositions could then be envisage, like a partial profit redistribution all along the supply chain.

4   Conclusions

This article presents a brief overview of existing EOL data management approaches proposed by the HMS community. Methods, techniques and materials are available to develop EOL management systems. However, some business limitations seem to slow down the adoption of these solutions. This is currently changing because investors are more and more sensitive to ecological questions, and also because service economy has been proved to be a credible economical alternative. Future research in EOL data management should focus on research axis around data synchronisation/dissemination/aggregation techniques and product-driven decision making applied on the EOL process. Demonstrating the performances and ROIs of such systems is a also key for their wide adoption.

References

1. Sempels, C., Hoffman, J.: Les business models du futur. Pearson (2013). ISBN: 978-2-3260-0026-1
2. Govindan, K., Soleimani, H., Kannan, D.: Reverse logistics and closed-loop supply chain: A comprehensive review to explore the future. Eur. J. Oper. Res. **240**(3)603–626 (2015). (Online). Available: http://www.sciencedirect.com/science/article/pii/S0377221714005633
3. Parlikad, A., McFarlane, D.: RFID-based product information in end-of-life decision making. Control Eng. Pract. **15**(11)1348–1363 (2007)
4. Le Moigne, R.: L'Économie circulaire. Dunod, France (2014). ISBN: 978-2-10-06008-3
5. Bentaha, M.L., Battaïa, O., Dolgui, A., Hu, S.J.: Dealing with uncertainty in disassembly line design. CIRP Ann. Manuf. Technol. **63**(1)21–24 (2014)
6. McFarlane, D., Sheffi, Y.: The impact of automatic identification on supply chain operations. Int.J. logistics Manage. **14**(1), 1–17 (2003)
7. Kiritsis, D.: Closed-loop PLM for intelligent products in the era of the Internet of Things. Comput. Aided Des. **43**(5)479–501 (2011)
8. Meyer, G.G., Främling, K., Holmström, J.: Intelligent products: A survey. Comput. Ind. **60**(3), 137–148 (2009). (Intelligent Products). (Online). Available: http://www.sciencedirect.com/science/article/B6V2D-4VCNDW2-1/2/8d4e089750b92f69fdff42cc12268818

9. Kubler, S., Derigent, W., Främling, K., Thomas, A., Rondeau É.: Enhanced product lifecycle information management using communicating material. Comput.Aided Des. **59**, 192–200 (2015)
10. Kubler, S., Främling, K., Derigent, W.: P2P Data synchronization for product lifecycle management. Comput. Ind. **66**, 82–98 (2015)
11. Koestler, A.: The ghost in the machine. Hutchinson (1967)
12. Van Brussel, H., Wyns, J., Valckenaers, P. Bongaerts, L., Peeters, P.: Reference architecture for holonic manufacturing systems: Prosa. Comput. Ind. **37**(3), 255–274 (1998). (Online). Available: http://www.sciencedirect.com/science/article/B6V2D-3V73RSY-7/2/30a8063959 d379b3fce4e70c664aa3ab
13. Brock, D.: The electronic product code (epc)—a naming scheme for physical objects. MIT Auto-ID Center White Paper, Jan 2001
14. Kärkkäinen, M.: Increasing efficiency in the supply chain for short shelf life goods using RFID tagging. Int. J. Retail Distrib. Manage. **31**(10), 529–536 (2003)
15. Fuentealba, C., Simon, C., Choffel, D., Charpentier, P., Masson, D.: Wood products identification by internal characteristics readings. In Industrial Technology. IEEE ICIT'04. IEEE International Conference on, vol. 2. IEEE, pp. 763–768 (2004)
16. Jover, J., Thomas, A., Leban, J.M., Canet, D.: Interest of new communicating material paradigm: An attempt in wood industry. J. Phys: Conf. Ser. **416**(1), 012031 (2013)
17. DIALOG, Distributed information architectures for collaborative logistics. Available from: http://dialog.hut.fi (2009). Accessed 14 Nov 2009. Technical Report
18. Kahn, O., Scotti, A., Leverano, A., Bonini, F., Ruggiero, G., Dörsch, C.: Rfid in automotive: a closed-loop approach. In: Proceedings of ICE, vol. 6 (2006)
19. Ranasinghe, D., Harrison, M., Främling, K., McFarlane, D.: Enabling through life product-instance management: solutions and challenges. J. Netw. Comput. Appl. (2011)
20. T. O. Group, O-MI, Open messaging interface, an open group internet of things (IoT) standard, Reference C14B, US ISBN 1-937218-60-7, Std., Oct 2014
21. T. O. Group, O-DF, Open data format, an open group internet of things (IoT) standard, C14A, US ISBN 1-937218-59-1, Std., Oct 2014
22. Parlikad, A., McFarlane, D.C., Fleich, E., Gross, S.: The role of product identity in end-of-life decision making. Auto ID Centre White Paper CAM-AUTOID-WH017. Technical Report (2003)
23. Vegetti, M., Leone, H., Henning, G.: Pronto: An ontology for comprehensive and consistent representation of product information. Eng. Appl. Artif. Intell. 24(8)1305–1327(2011) (semantic-based Information and Engineering Systems). (Online). Available: http://www. sciencedirect.com/science/article/pii/S0952197611000388
24. Rachuri, S., Subrahmanian, E., Bouras, A., Fenves, S.J., Foufou, S., Sriram, R.D.: Information sharing and exchange in the context of product lifecycle management: role of standards. Comput. Aided Des. **40**(7), 789–800(2008)
25. Harrison, M.: The 'internet of things' and commerce. XRDS: Crossroads ACM Mag. Students. **17**(3), 19–22(2011)
26. MIMOSA. (Online). Available: www.mimosa.org
27. Derigent, W.: Aggregation of product information for composite intelligent products. In: 9th International Conference on Modeling, Optimization & SIMulation (2012)
28. Mekki, K., Derigent, W., Zouinkhi, A., Rondeau, E., Abdelkrim, M.N.: Data dissemination algorithms for communicating materials using wireless sensor networks. In: International Conference on Future Internet of Things and Cloud (FiCloud), IEEE. pp. 230–237 (2014)
29. Mekki, K., Derigent, W., Zouinkhi, A., Rondeau, E., Thomas, A., Abdelkrim, M.N.: Non-localized and localized data storage in large-scale communicating materials: probabilistic and hop-counter approaches. Comput. Stan. Interfaces. (2015)
30. Pochampally, K.K., Gupta, S.M., Govindan, K.: Metrics for performance measurement of a reverse/closed-loop supply chain. Int. J. Bus. Perform. Supply Chain Model. **1**(1), 8–32(2009)

31. Ilgin.,M.A., Gupta,S.M.: Performance improvement potential of sensor embedded products in environmental supply chains. Resour. Conser. Recycl. 55(6), 580–592(2011). (Environmental Supply Chain Management). (Online). Available: http://www.sciencedirect.com/science/article/pii/S0921344910001126
32. Zhou, L., Disney, S.M.: Bullwhip and inventory variance in a closed loop supply chain. OR Spectr. **28**(1), 127–149(2006)

Part II
Recent Advances in Control for Physical Internet and Interconnected Logistics

The Internet of Things Applied to the Automotive Sector: A Unified Intelligent Transport System Approach

Valentín Cañas, Andrés García, Jesús Blanco
and Javier de las Morenas

Abstract This paper proposes a hybrid ITS design, which represents an adaptation of the Internet of Things to the automotive and transport fields. This proposal opens a large variety of new applications intended to improve significantly traffic safety, efficiency and organization. Furthermore, two completely new ideas are implemented: a new technology integration based on the Health and Usage Monitoring Systems philosophy which leads to a better diagnosis platform, and a slight modification of our own proposal in order to show a solution to avoid the use of multiple redundant ITS since they partially share both targets and technologies.

Keywords Intelligent transport systems · Internet of things · Vehicle to vehicle · Vehicle to infrastructure · Health and usage monitoring system

1 Introduction

Intelligent Transport Systems (ITS) are becoming a technological revolution in the transportation and automotive sector [1, 2]. The main goal of ITS is interconnecting all vehicles in a network so that safety and efficiency measures can be developed. Besides, an ITS system can offer more additional services. In fact, nowadays these

V. Cañas (✉) · A. García · J. Blanco
AutoLog Group, School of Industrial Engineering, University of Castilla-La Mancha,
Ciudad Real, Spain
e-mail: Valentin.Canas@Uclm.es

A. García
e-mail: andres.garcia@uclm.es

J. Blanco
e-mail: jesus.blanco@uclm.es

J. de las Morenas
AutoLog Group, Mining and Industrial Engineering, School of Almadén,
University of Castilla-La Mancha, Almadén, Spain
e-mail: javier.delasmorenas@uclm.es

© Springer International Publishing Switzerland 2016
T. Borangiu et al. (eds.), *Service Orientation in Holonic and Multi-Agent
Manufacturing*, Studies in Computational Intelligence 640,
DOI 10.1007/978-3-319-30337-6_5

technologies are evolving into an adaptation of the Internet of Things (IoT) to the automotive field, which is emerging as one of the most important technological trends over the incoming years. The most representative implementations that have recently emerged follow two different philosophies.

On one hand, it is possible to find solutions based on Vehicle to Infrastructure (V2I) and Infrastructure to Vehicle (I2V) types [3, 4] which usually appear together. In this first kind of structure, each vehicle establishes an independent communication with the servers of the traffic operator through mobile network technology, as well as through electronic beacons placed at strategic points on the road using IEEE 802.11b/g/n interfaces. Therefore, the operator is responsible for coordinating all information exchanges. This type of solutions shows the advantage of providing coverage to the entire network of vehicles; however the high latencies of mobile networks at present—about 100 ms for 3G connections and 50 ms for 4G [5]—do still make it a solution that cannot be considered safe enough for the transport environment where many processes require lower latencies. In the near future however, 5G technology is destined to change this situation by offering less than 1 ms latency. Another problem of this solution comes from the fact of being conceived as highly centralized system, for which the computational power required in traffic operators' servers is considerably high [3].

An example of centralized ITS platform is derived from the new standard approved by the European Parliament about smart digital tachographs, published in the Official Journal of the European Union on February 28, 2014—Regulation (EU) No. 165/2014 [6]. Consequently, transport companies will be required to install smart tachographs in all their new vehicles from 2017. This regulation requires new tachographs to incorporate a mobile data connection, GPS and speed sensors, so that they can provide permanent access to the authorities, leading to a V2I/I2V basic system.

The other ITS kind of solutions follows a Vehicle to Vehicle (V2V) morphology, where a fully decentralized model, in which all communications are inter-vehicular, arises. These communications between vehicles are established taking advantage of the IEEE 802.11p [2] specification, especially suitable for inter-vehicular data transfers due to its range and low latencies, which ensure that critical information can flow safely. Furthermore, decentralization fosters that the required computational load is distributed between each of the vehicles, forming a network of intelligent nodes interacting as a multi-agent system. By contrast, a more orthodox centralized implementation shows difficulty exchanging information with isolated vehicles. The decentralized idiosyncrasy and the limited range of the IEEE 802.11p interface may lead to isolated islands temporarily created through the traffic flow.

As follows from the above, the application cases for which the V2I/I2V structure is more suitable and relevant are those where V2V losses ground and vice versa, which makes these technologies complementary. This is why finally a hybrid solution has been proposed in recent years, the V2X morphology where both V2I/I2V and V2V are combined, leading to a decentralized network in which an operator supervises the traffic [7–9]. Derived from that idea, one of the major contributions of this paper is the design of a modular system with the capability to operate simultaneously as a V2X type ITS system and as a smart digital tachograph according to the above regulations.

2 ITS System Proposal

Implementing an ITS capable of integrating a large number of different types of vehicles, while ensuring that privacy and safety are guaranteed for all components, represents a major challenge. The proposed solution supports both a V2X ITS network and the smart tachograph systems, given the modularity of the design, which lead to great advantages in terms of lower costs and resources optimization.

Figure 1 shows the morphology corresponding to the proposed V2X ITS design. Communications between vehicles (V2V) are performed by the IEEE 802.11p interface. The vehicles also maintain a direct communication V2I/I2V with the cloud through mobile network connections, where the technology to be applied— 2G, 3G, 4G—is decided depending on availability. The ITS platform servers include applications that are enabled to the different stakeholders: authorities— police and traffic operators—, roadside services, car manufacturers, transport companies, mobile and web applications for drivers, etc. Traffic operators also maintain remote control over traffic lights, signals and electronic panels through this same system. Besides, these devices interact as electronic beacons with the vehicles, so that they can represent an additional I2V communication channel in addition to the mobile network.

In this solution, the modularity of ITS vehicle units is exploited to incorporate a system that adds diagnostic capabilities and forecast status of various vehicle components. Therefore, the unit collects information from two types of sources:

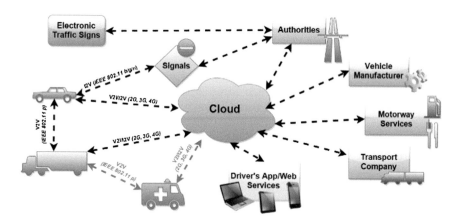

Fig. 1 Proposed V2X ITS system

accelerometers arranged at critical points of the vehicle on which an analysis using techniques based on Fourier transform is performed applying the HUMS techniques, as well as reading diagnostic messages on the engine's data bus.

3   Services Provided by the Proposed System

The main motivation of the ITS is to increase safety on the roads. Consequently, a large amount of applications are focused on reducing road accidents, which in other words means saving lives. Due to the fact that in most of the accidents more than one car is involved, and also because of the short lapse of time accidents occur, it is critical to use a proper technology able to carry out such applications. This results in the choice of the new IEEE 802.11p standard, which is focused on V2V communications and provides very low latencies. Some applications enabled by this technology are alert of collision risk, focused on both intersection and frontal collisions, the monitoring of the distance with following car and consequent alerts and automatic braking.

In contrast, other applications are better performed using mobile data communications. As examples, the presence of emergency vehicle alerts, slow, stopped or even wrong way incoming vehicles, or recent accidents as well, can be notified some kilometres before reaching the corresponding location.

This concept also fits with the alert of traffic jams and roadside works, which ties in with other family of applications: those related to improving traffic efficiency. Therefore, the system analyses the road conditions and proposes alternatives to the current one. Another application is that called 'green light countdown'. In this case, when approaching a traffic light showing red, the HMI shows a countdown for the status change from red to green. This improves fuel consumption as the driver can reduce the speed in order to avoid stopping the vehicle. Another application consists on a diagnosis management platform for the maintenance of vehicle parts, which is carried out applying HUMS techniques.

The diagnosis platform is related to efficiency, as it promotes an improvement of resources usage and engines efficiency, but it can be also considered as a service for drivers. In fact, the proposed ITS supports several applications related to services. For example, petrol stations—considering fuel consumption—and roadside restaurants are shown along the route focusing on most outstanding deals. The proposed system also analyses fuel consumption, proposing changes in driving in order to enhance the obtained profiles.

This last service can also be used by transport companies as a tool for improving fuel consumption in their fleets. Hence, it can be also considered as a logistic application.

However, the main logistic application consists on a fleet's management platform which has the capability of incorporating repair, refuelling and catering services so that the system can organize and coordinate mandatory breaks with refuelling, inspection and repairs. Besides, the platform behaves as a commander centre able to have a real time monitoring of the whole fleet.

4   ITS Unit Design

ITS units installed in each of the vehicles that are part of the ITS network are the main devices enabling system operation as a whole.

In the case of standard vehicles—such as cars—the requirements for an ITS unit can be classified into six families: telematic networks, diagnosis and prognostics system, driver interface, vehicle interface, security and computing power to manage ITS services and applications. The ITS unit also has a battery that guarantees the operation of the unit even when the vehicle is turned off.

This allows the detection of possible accidents and the emergency call service to be enabled in critical situations.

As shown in Fig. 2, the design for standard vehicles focuses on three main units. These are the Telematic Control Unit (TCU) which manage telematic networks, the HUMS Diagnosis Unit (HDU) which manage the diagnosis system, and the Application Unit (AU). This coordinates all the ITS unit modules and is responsible of the management of the ITS services as well as the interfaces between the ITS system and both the vehicle—through electronic units bus data—and the driver—through a HMI provided with a screen and speakers.

The TCU unit, whose design is shown in Fig. 3, has wireless connection including a variety of supported technologies such as: IEEE 802.11p for communication between vehicles, IEEE 802.11b/g/n for connections with signalling beacons, and mobile networks—2G, 3G and 4G—for real time data transfer with ITS platform servers.

It also has a GPS module to obtain the vehicle position and a Bluetooth module that is used to refresh the system data to the mobile app. A TCP/IP based connection maintains a local network between the three main units—TCU, HDU and AU. The TCU and HDU have the capability of reading the vehicle electronic units' data bus—CAN, LIN, FlexRay or Ethernet—in order to minimize latency when errors are reported.

The HDU unit, as shown in Fig. 4, is responsible for processing the signals received from various sensors distributed throughout the as well as from shared

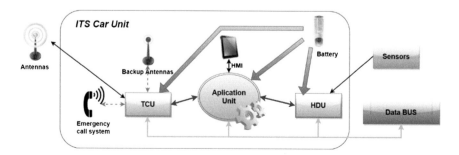

Fig. 2 ITS unit design for standard vehicles

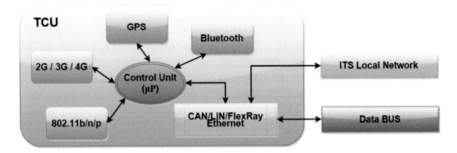

Fig. 3 TCU design

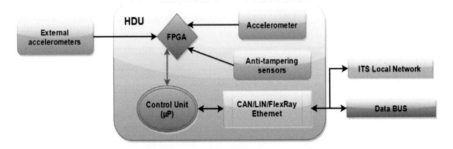

Fig. 4 HDU design

variables via bus communication with vehicle's electronic units. This processing is carried out at two levels of intelligence: instant intelligence—diagnosis—and predictive intelligence—prognosis.

5 Smart Tachograph Integration as an ITS Unit

One of our purposes in this paper is to provide a modular design capable to incorporate the smart tachograph in industrial vehicles as an ITS unit. This solution provides a more complex and useful V2X system than what has been proposed in the new EU legislation, so that a common ITS can be set-up among all vehicles in a road. Thus, the design of a digital tachograph would be merged with that of a prototype ITS V2X providing extended functionality.

The new device has all necessary interfaces for V2X system networks, covering all network requirements for operating as a digital tachograph. Furthermore, the tachograph's HMI interface is integrated with the ITS one. The unification of the two embedded designs also helps to improve the designs of prototype ITS V2X units for standard private vehicles, since security measures would be reinforced as a

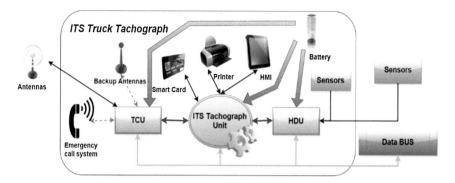

Fig. 5 Proposed design for a smart tachograph integrated in the ITS unit

result of the expected spreading in the implementation of anti-hacking strategies that are commonly used in digital tachographs.

Additionally, thanks to the proposed modular design, the most important components will be able to address a broader market than only that of industrial vehicles. Therefore this results in a more efficient use of resources as the proposed design simplifies the smart tachograph and the ITS standard unit in a single generic device that avoids inefficient duplication of components. By economy of scale, in addition, manufacturers will thus be able to reduce costs.

The combination of a digital tachograph without network interfaces, with the additional features of a TCU unit and a HDU unit, results in a tachograph that adapts to the new EU legislation, while supporting the ITS V2X morphology proposed in this paper.

As shown in Fig. 5, the major changes in the design of the unit as tachograph is the existence of an ITS Tachograph Unit instead of the Application Unit used in the generic case. This unit is responsible to perform the same operations as the Application Unit, but also supports the services necessary for the operation of the platform as a digital tachograph. Other important changes include the presence of a smart card slot and a printer, both required by law, as well as additional built-in sensors to prevent hacking.

6 Conclusions and Future Research

In this paper the current status of ITS has been reviewed. Additionally, an ITS based on V2X architecture has been proposed, adding two completely new ideas.

The utilisation of the HUMS philosophy to improve parts fault detection and other incidents in civil transport can be highlighted as one of the main contributions of the proposed models.

Furthermore, this work lays the foundations for a common platform that integrates new smart tachographs in industrial vehicles with an ITS, and for which a first prototype has been developed.

Future research will focus on studying the behaviour of the improved prototype units, now under development, as well as of the related software and apps. It arises as especially relevant the setting-up of complete system simulators, that would include the implementation of banks of servers as well as the development of the applications required to provide additional services to all participants in the proposed system. Once these first complete systems are in place, special attention will have to be paid to deepen the study of the system security.

Acknowledgements This work is supported by the resolution of 31/07/2014, published by the University of Castilla-La Mancha, which establishes the regulatory bases and the call for predoctoral contracts for the training of researchers under the Plan Propio de I+D+i, co-financed by the European Social Fund. [2014/10340].

References

1. Dressler, F., Hartenstein, H., Altintas, O., Tonguz, O.K.: Inter-vehicle communication: Quo Vadis. IEEE Commun. Mag. **52**, 170–177 (2014)
2. Milanes, V., Onieva, E., Perez, J., Simo, J., Gonzalez, C., de Pedro, T.: Making transport safer: V2V-based automated emergency braking system. Transport **26**, 290–302 (2011)
3. Godoy, J., Milanes, V., Perez, J., Villagra, J., Onieva, E.: An auxiliary V2I network for road transport and dynamic environments. Transp. Res. Part C-Emerg. Technol. **37**, 145–156 (2013)
4. Milanes, V., Villagra, J., Godoy, J., Simo, J., Perez, J., Onieva, E.: An intelligent V2I-based traffic management system. IEEE Trans. Intell. Transp. Syst. **13**, 49–58 (2012)
5. Feteiha, M.F., Hassanein, H.S.: Enabling cooperative relaying VANET clouds over LTE-A networks. Veh. Technol. IEEE Trans. on **64**, 1468–1479 (2015)
6. Regulation (EU) No 165/2014 of the European Parliament and of the Council of 4 February 2014 on tachographs in road transport, repealing Council Regulation (EEC) No 3821/85 on recording equipment in road transport and amending Regulation (EC) No 561/2006 of the European Parliament and of the Council on the harmonisation of certain social legislation relating to road transport Text with EEA relevance. OJ L 60, 28.2.2014, pp. 1–33 (BG, ES, CS, DA, DE, ET, EL, EN, FR, GA, HR, IT, LV, LT, HU, MT, NL, PL, PT, RO, SK, SL, FI, SV)
7. Wiesbeck, W., Reichardt, L.: C2X communications overview, Electromagnetic Theory (EMTS), 2010 URSI International Symposium on, pp. 868–871 (2010)
8. Barrachina, J., Sanguesa, J.A., Fogue, M., Garrido, P., Martinez, F.J., Cano, J.C., Calafate, C. T., Manzoni, P.: V2X-d: A vehicular density estimation system that combines V2V and V2I communications, Wireless Days (WD), 2013 IFIP, pp. 1–6 (2013)
9. Parrado, N., Donoso, Y.: Congestion based mechanism for route discovery in a V2I-V2V system applying smart devices and IoT. Sensors **15**, 7768–7806 (2015)

Using the Crowd of Taxis to Last Mile Delivery in E-Commerce: a methodological research

Chao Chen and Shenle Pan

Abstract Crowdsourcing is gathering increased attention in freight transport areas, mainly applied in internet-based services to city logistics. However, scientific research, especially methodology for application is still rare in the literature. This paper aims to fill this gap and proposes a methodological approach of applying crowdsourcing solution to Last Mile Delivery in E-commerce environment. The proposed solution is based on taxi fleet in city and a transport network composed by road network and customer self-pickup facilities that are 24/7 shops in city, named as *TaxiCrowdShipping* system. The system relies on a two-phase decision model, first *offline taxi trajectory mining* and second *online package routing and taxi scheduling*. Being the first stage of our study, this paper introduces the framework of the system and the decision model development. Some expected results and research perspectives are also discussed.

Keywords Last mile delivery · Crowdsourcing · Taxi trajectory data mining · Freight transport · City logistics

1 Introduction

In E-commerce environment, Last Mile Delivery (hereinafter LMD) is the problem of transport planning for delivering goods from e-retailers' hubs to the final destination in the area, for example the end consumers' homes, see [1] and [2]. Speed and cost are the two crucial success factors to LMD. Faster shipping while with

C. Chen
College of Computer Science, Chongqing University,
144, Shazheng Street, Chongqing, China
e-mail: cschaochen@cqu.edu.cn

S. Pan (✉)
Centre de Gestion Scientifique - I3 - UMR CNRS 9217,
MINES ParisTech - PSL Research University, 60, Bd St Michel, Paris, France
e-mail: shenle.pan@mines-paristech.fr

© Springer International Publishing Switzerland 2016
T. Borangiu et al. (eds.), *Service Orientation in Holonic and Multi-Agent Manufacturing*, Studies in Computational Intelligence 640,
DOI 10.1007/978-3-319-30337-6_6

61

lower cost is the major challenge; nevertheless, it is also a paradox to a certain extend. Indeed, when customers are given a choice between fast and cheap delivery, most of them choose the cheap one, as observed by a recent report [3]. The report also infers that low-cost, speedy two-day delivery corresponds to most customers' expectation, opposite to the one-day delivery policy pursued by giant e-retailers such as Amazon and Alibaba, etc. This fact may open up new opportunities to innovative freight transport models [4] for LMD aiming at reducing delivery cost while respecting shipping time, nevertheless, not necessarily aiming at minimizing shipping time. Being our topic here, crowdsourcing is one of such solutions getting more and more attention [5].

In the literature, crowdsourcing has been usually seen as "an interesting idea" for freight transport before seriously moving to real applications. Despite the existence of some internet-based services, scientific research, especially methodology for application is rare in the literature [5]. This paper aims to fill this gap by providing a methodological approach of applying the crowdsourcing solution proposed and to assess its performance. In this paper, the crowd studied is taxi fleet in city, supported by a transport network composed by road network and customer self-pickup facilities such as 24 h shops in city, named as *TaxiCrowdShipping* system. The system relies on a two-phase decision model, first offline taxi trajectory mining and second online package routing—taxi scheduling. As the first stage of our research, this paper introduces the objective and the framework of the *TaxiCrowdShipping* system, as well as to define the function of system.

The reminder of the paper is organised as follows. Section 2 consists of a relevant literature review. Then, Sect. 3 presents the *TaxiCrowdShipping* system. After introducing some basic concepts and assumptions, we focus on the 2-step decision support tool for the system. Some expected results are also discussed. Finally, Sect. 4 concludes this paper by giving some research perspectives for future works.

2 Related Works

Recently some innovative solutions have been studied for city logistics and LMD in E-commerce environment, for example those involved in our study like interconnected city logistics enabled by Physical Internet [6, 7], self-service parcel station (e.g., DHL PackStation, LaPoste Pickup Station etc.), new tools for LMD (bicycle, motor, electric vehicle etc.), Smart city logistics [8], and crowd sourced delivery [5]. Due to space limitation, here we focus on the works on crowdsourcing in Freight Transport.

Being firstly discussed in [9], crowdsourcing has been increasingly studied as solution to freight transport. It can be simply defined as "*outsourcing a task to the crowd* via *open call*" [9]. On the practice side, it occurs mainly in the form of internet-based services for example *imoveit.co.uk* and *zipments.com*, where the crowd is undefined. Thus, both professional (e.g., carriers) and non-professional (e.g., inhabitants) service providers may answer the calls. In 2014 Amazon has launched a project to explore taxi deliveries in San Francisco and Los Angeles.[1] The idea is similar to our study, though, their methodology and results are not yet published to our knowledge. Moreover, the package deliveries are completed by ordering free taxis, while our proposed solution leverages the *hitchhiking rides* provided by occupied taxis when they are sending passengers; thus our solution is more green and economic. On the scientific side, only few relevant works in the area of logistics can be found. A case study of applying crowdsourcing to library deliveries in Finland is conducted in [5]. They study a system called *PiggyBaggy* to assess the sustainability and adaptability of such solution. A taxi-based solution for the waste-collecting or product return problem (i.e., reverse logistics) in metropolitan area is discussed in [10], without considering goods delivery. Some other relevant works can be also found in the area of data science. Data scientists are mainly interested at mining the taxi trajectory data to understand the city dynamics, and developing various smart services for taxi drivers, passengers, as well as the city planners [11, 12]. However, almost all the current research related to taxi data mining focuses on the people or public transport [13, 14]; little attention has been paid to freight transport.

From the literature we can see that crowdsourcing in freight transport usually occurs in the form of internet-based services in practice, and it is usually investigated via case study in the literature. Methodology for application is not well addressed. Besides, no attention has been paid to crowd selection or definition. People in city are often regarded as eligible crowd. Following the previous work [10] dealing with reverse flows, this paper focus on a methodology approach for the LMD problem, where logistics constraints and decision model are different.

3 *TaxiCrowdShipping* System

3.1 *Basic Concepts and Assumptions*

To ease the description, we define the related concepts based on Fig. 1, and also make some assumptions.

Definition 1 (*Road Network*) A road network is a graph *G(N, E)*, consisting of a node set *N* and an edge set *E* (as shown in Fig. 1), where each element *n* in *N* is an intersection and is associated with a pair of longitude and latitude degrees *(x, y)* representing its spatial location. Edge set *E* is a subset of the cross product *NxN*. Each element *e(u, v)* in *E* is a street connects node *u* to node *v*, which can be one-way or bi-directional, depending on real cases.

---

[1]http://www.engadget.com/2014/11/05/amazon-is-exploring-taxi-deliveries-in-san-francisco-and-los-ang/.

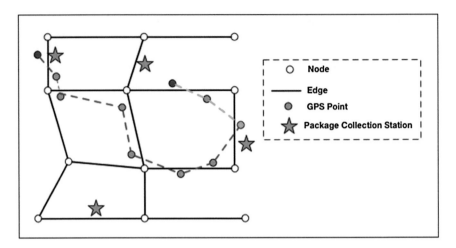

Fig. 1 Illustration of some basic concepts

Definition 2 *(Taxi Trajectory)* A taxi trajectory is a sequence of time-stamped GPS points. Each GPS point $p_i = (t_i, x_i, y_i, ind_i)$ consists of a time-stamp t_i, a longitude x_i, a latitude y_i, and an indicator ind_i showing whether the taxi is occupied or not. A pickup point is a special GPS point, with the indicator changing from 0 to 1 (the red-coloured circle in Fig. 1), while a drop-off point is the one with the indicator changing from 1 to 0 (the green-coloured circle in Fig. 1). Thus we can further define a *passenger-delivery trajectory* is the GPS sequence from the pick-up point to the followed drop-off one (red dashed line in Fig. 1); a *passenger-hunting trajectory* is the GPS sequence from the drop-off point to the next pickup point (green dashed line, Fig. 1).

Definition 3 *(Package Pickup Station)* A package pickup station is a Point of Interest (POI hereinafter) near roadside that is responsible for storing packages waiting for consumer pickup, (the star in Fig. 1). Here we select *24-h opening convenience stores* near roadside as the package pickup stations.

Definition 4 *(Package Delivery Request)* A package delivery request is defined as a triple $<o_p, d_p, t_p>$, where o_p and d_p refer to the origin and the destination of the package respectively, and t_p refers to the time when the user submits the request. The request is generated by users who need the package express delivery service.

Definition 5 *(Real-time Taxi Ordering Request)* A real-time taxi ordering request is defined as a triple $<o_t, d_t, t_t>$, where o_t and d_t refer to the passenger's origin and the destination respectively; t_t refers to the time when the passenger submits the request. The request is made by passengers who need taxi service.

Assumption 1 All selected POI is open 24/7, without capacity issue and with good accessibility to taxi.

Assumption 2 Taxi drivers are willing to accept assigned package delivery tasks.

Assumption 3 The package can be *trackable*. Since the birth time, the package is either stored at the pickup station or carried by the scheduled taxi. Each pickup station is authorized and has a unique ID; each taxi is registered in taxi management department and also has a unique ID.

3.2 Problem Description

To help understand how our proposed solution works to handle the LMD, we intentionally design a simple running example. Suppose in Fig. 1 the leftmost and rightmost stars are the origin and destination of the package respectively. After generating package delivery request, there happens to be a passenger who makes a real-time taxi ordering request, intending to go to the same destination. At that time, we can assign the package delivery task to the taxi which has accepted the passenger's request. Finally, the package will be also delivered, with a hitchhiking ride provided by the taxi while sending the passenger. The solution can be featured as economic and eco-friendly since it almost does not incur extra labour cost, energy and CO_2 emissions.

Accordingly, the taxi-based crowdsourcing solution to LMD consists of the shortest path finding problem for packages and the scheduling problem for taxis, and it can be described as follows.

Given

- A road network and a set of package pickup stations in the studied city;
- A set of taxi trajectory data in the studied city in history (e.g., last month);
- A set of package delivery requests, and a set of real-time generated taxi ordering requests. Note that these requests come in stream.

Objective

For a given package delivery request, find its optimal delivery path which can minimize the total package delivery time (i.e. maximize the delivery speed). Once determined the path, we can schedule the next coming taxi to deliver the package having the same destination. Note that in one of the scenarios in this study the path can be re-planned according to the Real-time Taxi Ordering Request.

Constraints

Only taxis which respond the taxi ordering requests after the package delivery request can be scheduled. Once a taxi is involved into a delivery task, it can be available again to be scheduled to participate only after completing the current task (i.e. sending the package to the predefined pickup station). In other words, a taxi can carry at most one package when sending passengers.

3.3   *The* TaxiCrowdShipping *System*

The *TaxiCrowdShipping* system contains two components, i.e., the *Offline-Trajectory Mining and the Online-Package Routing* respectively, which will be detailed as follows and shown in Fig. 2.

3.3.1   Offline-Trajectory Mining

The objective of the Offline-Trajectory Mining is to estimate the direct package delivery time from one pickup station to another, by taking a single hitchhiking ride. The time cost mainly includes two parts: the time cost on waiting for the hitchhiking rides, which is related to the frequency of taxi rides, and the time spent on driving on the roads. Here, we propose a two-step procedure to estimate the time cost.

Step 1:   *From Trajectory Data to Passenger Flow.* From the given taxi trajectory data it is not difficult to compute the passenger flow between any two pickup stations. Specifically, to compute the passenger flow from cs_i to cs_j, the trajectories meeting Eqs. (1)–(2) will be counted. Then, the passenger flow from cs_i to cs_j is just the number of trajectories satisfying the requirements. Note that there may be no passenger flow between some pickup station pairs, and the passenger flow is different at different time slots. For the purpose of future research, we divide the time into three time slots for a day in advance, i.e., night-time hours, day-time hours and rush hours

$$Ddist(Tr_i.o, loc(cs_i)) < \delta \qquad (1)$$

$$Ddist(Tr_i.d, loc(cs_j)) < \delta \qquad (2)$$

where $Tr_i.o$ and $Tr_i.d$ are the original and destination points of Tr_i, respectively; $loc(\cdot)$ gets the latitude and longitude location of the given pickup station; δ is a user-specified parameter. $Ddist(a \cdot b)$ calculates the driving distance from point a to b.

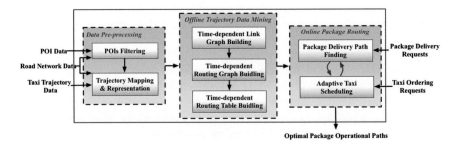

Fig. 2 The *TaxiCrowdShipping* system architecture

Step 2: *From Passenger Flow to Time Cost.* To estimate the time cost, we need to estimate two parts, i.e. the waiting time and the driving time. The waiting time is defined as the time cost on waiting for the suitable hitchhiking ride event of passenger taking taxis, to help deliver a package from cs_i to cs_j directly (with no transhipment). Here, we employ the *Non-Homogeneous Poisson Process (NHPP)* to model the behaviour of passenger taking taxis [15]. According to the passenger flow, we can estimate the waiting time of packages at different time slots at the pickup stations. Under the Poisson hypothesis within a time slot, we can derive the probability distribution of the waiting time for the next suitable hitchhiking ride event (i.e. t_{next}, the event of a passenger taking taxi from cs_i to cs_j), which can be expressed in Eq. 3:

$$\begin{aligned} P\{t_{next} \leq t\} &= 1 - P\{t_{next} > t\} \\ &= 1 - P\{N(t) = 0\} \\ &= 1 - e^{-\lambda \cdot t} \end{aligned} \tag{3}$$

Here $N(t)$ is the number of event occurring in t, and $P\{'N'(t) = k\} = e^{-\lambda \cdot t} \frac{(\lambda \cdot t)^k}{k!}$ Then the probability density function (pdf) of t_{next} is the derived function of $P\{\}$, Eq. 4

$$p(t) = \lambda \cdot e^{-\lambda \cdot t} \tag{4}$$

Thus, we can deduce the expectation of t_{next} (i.e. the waiting time for the hitchhiking ride event occurring):

$$E[t_{next}] = \int_0^\infty t \cdot \lambda \cdot e^{-\lambda \cdot t} \cdot dt = \frac{1}{\lambda} \tag{5}$$

Note that λ in the model is the frequency of passenger taking taxis from cs_i to cs_j (i.e. the passenger flow from cs_i to cs_j), which can be easily estimated by the Eq. 6.

$$\hat{\lambda} = \frac{\bar{N}}{\Delta T} \tag{6}$$

where $\bar{N}$ is the average number of passengers taking taxis from cs_i to cs_j during the studied time slot in the observed days; ΔT is the time duration of the that time shot. Therefore, the waiting time from cs_i to cs_j is:

$$waiting\ time = \frac{1}{\hat{\lambda}} = \frac{\Delta T}{\bar{N}} \tag{7}$$

For each passenger-delivery ride from cs_i to cs_j, it is easy to derive its time spent on driving on the roads. The driving time is the average one of all such rides, as in Eq. 8.

$$driving\ time = \frac{\sum_{i=1}^{N} Tr_i \cdot (te - ts)}{N} \tag{8}$$

where N is the number of passenger-delivery rides during the studied time slot in the observed days $t_e - t_s$ is the time cost of the corresponding taxi ride. Finally, the time cost is just the sum of waiting time and driving time, as in Eq. 9. Note that the time cost will be $+\infty$ if there was no passenger flow on that pickup station pair.

$$tc = waiting\ time + driving\ time = \frac{\Delta T}{N} + \frac{\sum_{i=1}^{N} Tr_i \cdot (te - ts)}{N} \tag{9}$$

3.3.2   Online-Package Routing and Taxi Scheduling

The objective of the Online-Package Routing and Taxi Scheduling is to schedule the specific taxis to help delivery the packages with the determined optimal path, according to the real-time coming taxi ordering requests. Here, we also propose a two-step procedure to complete, detailed as follows.

Step 1: *Find the Optimal Pickup Station Sequence.* For a package delivery request, with the estimated time cost values in the last component, it is trivial to find the best pickup station sequence from the origin to the destination of the package, in terms of the total time cost, by applying the classical shortest path finding algorithms

Step 2: *Schedule the Taxis.* After obtaining the optimal pickup station sequence for a package delivery request, we schedule the taxis according to the real-time taxi ordering requests. In more detail, from the origin of the package to the followed pickup station in the optimal pickup station sequence, we wait for the taxi which will pick up a passenger at the origin, heading to the followed station (the information is included in the real-time taxi ordering requests), and assign the package delivery task to that taxi. After that, the origin of the package will be also updated. The procedure will be repeated until the package arrives at its destination

3.4   Expected Results

Following the framework proposed here some results are expected in the next steps. First, we will conduct one study to assess the possibility to implement the *TaxiCrowdShipping* system proposed. A large city in China, namely Hangzhou city is selected to be the test field, thanks to some available data sets there such as Open data of taxi trajectory, may of city shops' and road network etc. However, the data of package delivery request is still to be completed. Second, a set of algorithms for package routing and taxi scheduling problem will be developed and examined. Then a set of scenarios will be run to assess the performance of the system and its sensibility.

4 Conclusion

In this paper we aim to propose a methodological approach of applying crowd-sourcing solution to Last Mile Delivery (LMD) problem in city logistics. To this end, a system called *TaxiCrowdShipping* is proposed, whose objective is to use the taxi fleet and shops in city for LMD purpose. The framework of such system is discussed, as well as the two-phase decision model. Except the expected results discussed above, this study opens some research perspectives. First, the Physical Internet-based active container described in [16] may be adapted to this study. Suppose that such containers are able to actively publish a delivery request on web (as a passenger calls a taxi), the taxi scheduling could be more efficient and responsive. It could also provide a good real time tracking and tracing technique to the system. Second, the study can be extended to automated self-service parcel station implantation problems. Coupling such station placement and crowdsourcing solution is still rarely studied in the literature.

References

1. Lee, H.L., Whang, S.: Winning the last mile of e-commerce. MIT Sloan Manage. Rev. **42**(4), 54–62 (2001)
2. Punakivi, M., Yrjölä, H., Holmstroem, J.: Solving the last mile issue: reception box or delivery box? Int. J. Phys. Distrib. Logistics Manage. **31**(6), 427–439 (2001)
3. Gibson, B.J., Defee, C.C., Ishfaq, R.: The state of the retail supply chain: Essential Findings of the Fifth Annual Report. RILA, Dallas, TX (2015)
4. Meyer-Larsen, N., Hauge, J.B., Hennig, A.-S.: LogisticsArena—A platform promoting innovation in logistics, in logistics management. Springer, pp. 505–516 (2015)
5. Paloheimo, H., Lettenmeier, M., Waris, H.: Transport reduction by crowdsourced deliveries— a library case in Finland. J. Cleaner Prod. (2015)
6. Sarraj, R., et al.: Analogies between Internet network and logistics service networks: challenges involved in the interconnection. J. Intell. Manuf. **25**(6), 1207–1219 (2014)
7. Crainic, T.G. Montreuil, B.: Physical Internet Enabled Interconnected City Logistics, in 1st International Physical Internet Conference, 2015, Québec City, Canada (2015)
8. Neirotti, P., et al.: Current trends in smart city initiatives: some stylised facts. Cities **38**, 25–36 (2014)
9. Howe, J.: The rise of crowdsourcing. Wired Mag. **14**(6), 1–4 (2006)
10. Pan, S., Chen, C., Zhong, R.Y.: A crowdsourcing solution to collect e-commerce reverse flows in metropolitan areas. In: 15th IFAC Symposium on Information Control Problems in Manufacturing INCOM 2015, Canada, Ottawa, Elsevier (2015)

11. Castro, P.S., et al.: From taxi GPS traces to social and community dynamics: A survey. ACM Comput. Surv. (CSUR) **46**(2), 17 (2013)
12. Zheng, Y.: Trajectory data mining: an overview. ACM Trans. on Intell. Syst. Technol. (TIST) **6**(3), 29 (2015)
13. Chao, C., et al.: B-Planner: planning bidirectional night bus routes using large-scale taxi GPS traces. IEEE Trans. on Intell. Transp. Syst. **15**(4), 1451–1465 (2014)
14. Liu, Y., et al.: Exploiting heterogeneous human mobility patterns for intelligent bus routing. In 2014 IEEE International Conference on Data Mining (ICDM) (2014)
15. Qi, G., et al.: How long a passenger waits for a vacant taxi–large-scale taxi trace min-ing for smart cities. In: 2013 IEEE and Internet of Things (iThings/CPSCom), IEEE International Conference on Cyber, Physical and Social Computing Green Comput. and Com. (GreenCom) (2013)
16. Sallez, Y., Pan, S., Montreuil, B., Berger, T., Ballot, E.: On the activeness of intelligent Physical Internet containers. Computers in Industry. http://dx.doi.org/10.1016/j.compind.2015.12.006. (2016)

Framework for Smart Containers in the Physical Internet

Ali Rahimi, Yves Sallez and Thierry Berger

Abstract In the context of the Physical Internet (PI), the PI-container with associated instrumentation (e.g., embedded communication, processing, identification…) can be considered as "smart". The concept of "Smart PI-container (SPIC)" exploits the idea for a container to participate in the decision making processes that concern itself or other PI-containers. This paper outlines the necessity to develop a framework able to describe a collective of SPICs. After a quick survey of the existing typologies in the field of smart entities, a descriptive framework based on an enrichment of the Meyer typology is proposed. The proposed framework allows a description of the physical aspect (links among PI-containers) and of the informational aspect for a given function. Finally, for illustration purpose, the framework is tested on a collective of SPICs for a monitoring application.

Keywords Smart PI-container (SPIC) · Framework · Intelligence · Physical internet · Logistics

1 Introduction

Montreuil [1] points out that current logistic systems are unsustainable economically, environmentally and socially. To reverse this situation, the author exploits the digital internet as a metaphor to develop an initiative called Physical Internet (PI). By analogy with data packets, the goods are encapsulated in modularly dimensioned,

A. Rahimi · Y. Sallez · T. Berger (✉)
Université Lille Nord de France, 59000 Lille, France
e-mail: thierry.berger@univ-valenciennes.fr

A. Rahimi
e-mail: ali.rahimi@univ-valenciennes.fr

Y. Sallez
e-mail: yves.sallez@univ-valenciennes.fr

A. Rahimi · Y. Sallez · T. Berger
UVHC, LAMIH UMR n°8201, 59313 Valenciennes, France

© Springer International Publishing Switzerland 2016   71
T. Borangiu et al. (eds.), *Service Orientation in Holonic and Multi-Agent Manufacturing*, Studies in Computational Intelligence 640,
DOI 10.1007/978-3-319-30337-6_7

reusable and smart containers, called PI-containers. This paper investigates the role of Smart PI-containers (addressed hereafter as SPIC) in the domain of Physical Internet. SPICs can take decision, and interact with other containers and actors of the PI network.

This paper proposes early work on the development of a descriptive framework permitting to analyse and classify different aspects of smart PI-containers. The Sect. 2 is dedicated to a presentation of the different categories of PI-containers and of the notion of SPIC. The requirements associated to the descriptive framework are then introduced. In Sect. 3, the existing typologies on smart entities are gathered and investigated, considering the previous requirements. The Sect. 4 describes the proposed framework and applies it on a collective of SPICs for monitoring application. Finally, conclusion and future perspectives are offered in Sect. 5.

2 Smart PI-Container (SPIC)

2.1 The Different Categories of PI-Containers

In the recent field of PI, current projects aim to refine the PI-container concept. As shown in Fig. 1, the LIBCHIP project [2] investigates the exploitation of three modular levels of PI-containers (and associated functionalities):

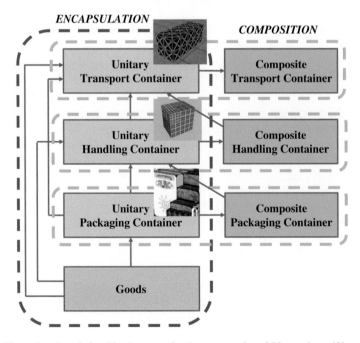

Fig. 1 Illustrating the relationships between the three categories of PI-containers [2]

- *Transport container*: Designed to be easily carried, to endure harsh external conditions and to be stackable as usual maritime shipping containers.
- *Handling container*: Designed to be easily handled by PI-handlers (conveying systems, lifts…) and to resist rough handling conditions.
- *Packaging container*: Designed to contain directly the physical goods. They basically replace the typical custom packaging, for example designed to market goods.
- The relationships between categories exploit two mechanisms (see Fig. 1):
- Encapsulation: The three categories can be successively encapsulated one within another.
- Composition: In a same category, the PI-containers can be composed and interlocked to build "composite" PI-containers and allow easier handling or transport, sharing the same standard type of interfacing devices.
- In this context, the European project MODULUSHCA [3] focuses on the design of handling PI-containers relying in modular construction and attachment between them. However, PI-containers must not only be considered as "standardized" containers with a cargo, but also "smart"; this is an important characteristic for PI management.

2.2   Smart PI-Containers

Through amplification of embedded communication and decisional capabilities, "smart" PI-containers can play an "active" role in PI management [4]. For example, the SPIC can be able to identify its state and send warnings when certain conditions are met (e.g. breaking the cold chain for perishable products). For more complex functions (e.g. routing of a SPIC in the PI network), a highest level of intelligence can be considered to adapt the current goals and negotiate with routing agents. According to Sect. 2.1, the intelligence of a SPIC can exploit interactions of a collective of PI-containers on three layers, as well as with the other PI means and agents. At each layer, SPIC can use different information sources: measurements on its own physical shell and skeleton (e.g. detection of shocks), on its inside physical environment (e.g. internal temperature) or on its external physical environment (e.g., temperature, interactions with other containers or with the Physical Internet management system).

To characterize a collective of SPICs in interaction, a descriptive framework is necessary. The associated requirements are presented in the next section.

2.3   Requirements for a Framework

In our context a framework is a conceptual structure intended to serve as "map" or guide useful for representing and analysing a collective of PI-containers. However, the "intelligence" associated to a SPIC must not be analysed as a whole but rather function by function. Indeed, from a decisional point of view, a SPIC can be "passive" for a function (e.g. f_i) and "active" for another one (e.g. f_j). Figure 2 highlights, for a specific PI-containers grouping and for a specific function f_i, five requirements that must be achieved by the framework according to physical and informational aspects:

- Physical aspect: (Req. 1) Which are the physical links existing between PI-containers in the collective of SPICs (i.e., encapsulation and composition)?
- Informational aspect: The four following requirements must be considered according to two points of view:

  - "Individual" point of view:
  - (Req. 2) *Intelligence level*: What is the intelligence level of each SPIC (e.g. from simple information handling to more complex decisional activities)?
  - (Req. 3) *Intelligence location*: How the intelligence of each SPIC is supported by a technology (i.e. embedded or remote implementations)?
  - "Collective" point of view:
  - (Req. 4) *Aggregation*: Which are the informational links in a hierarchy of SPICs (i.e. when several SPICs are included in one SPIC)?
  - (Req. 5) *Interactions*: Which are the interactions among SPICs (e.g. Master-Slave relationship)?

For a specific grouping of PI-containers the physical aspect is an invariant whatever are the function(s) considered. However the informal aspect evolves according to the studied function. In order to build the informational aspect of the framework, the next section offers a survey of the existing typologies in the field of Internet of Things, "smart" objects and "intelligent" products.

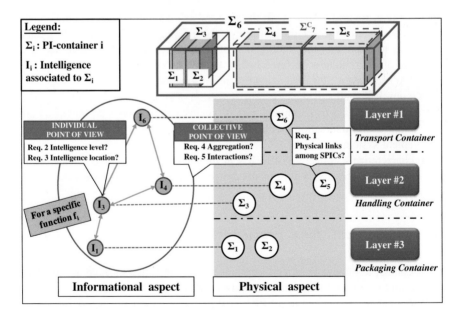

Fig. 2 Illustrating the requirements associated to the framework

3 Survey of Typologies on Smart Entities

Based on the study of Sallez [5], this section provides a brief survey of existing typologies. Two broad categories are distinguished: individual and collective.

3.1 Individual

This category focuses on the entity as "individual" and is in turn divided into two major classes: (i) *mono-criterion typologies* distinguishing broad classes of "intelligent" entities according to their level of intelligence, (ii) *multi-criteria typologies* taking into account the different characteristics of an "intelligent" entity (sensory capacities, location intelligence…).

Mono-criterion:
Le Moigne [6] proposed nine levels of intelligence, from a totally passive object at the first level to a self-completing active object at the highest level. Wong et al. [7] have proposed informational-oriented products and decisional-oriented products. Other typologies [8–10] have equally suggested different classifications of intelligence level focusing on different applications of smart entities.

Multi-criteria:
Meyer et al. [11] presented a typology based on three axes: level of intelligence, location of intelligence and aggregation of intelligence. Kawsar et al. [12] defined three sets of cooperating objects named SOS (Smart Object System) with five levels of intelligence. A three axis typology was introduced by Kortuem et al. [13], addressing awareness, interactivity and representation for smart objects. Three categories of smart *objects are then considered: activity-aware objects, policy-aware objects and* process-aware objects. López et al. [14] have proposed a five-level typology for smart objects starting from object identifying and storing all relevant data to finally object making decisions and participating in controlling other devices. In the same spirit, the typology of Sundmaeker et al. [15] proposes five categories of smart objects in the field of Internet of Things.

3.2 Collective

This category tries to characterize the types of interactions which exist in a collective of "intelligent" entities. The typology proposed by Salkham et al. [16] includes three aspects: goal, approaches and means including abilities as sensing and acting on the environment, communicating and delegating. In the field of

Internet of Things, the typology of Iera [17] was inspired by the theory of social relations of Fiske [18]. The four classes highlighted by Fiske are revisited in order to characterize the different relationships between entities.

3.3   Synthesis

Table 1 classifies the typologies according the informational requirements introduced in Sect. 2.3. In this table, a cross (+) is positioned when the typology considers explicitly the concerned requirement.

As highlighted by the Table 1, none of the typologies is sufficient to satisfy the four informational requirements. However, Meyer typology [11] exhibits the best score of all typologies. In the next section, this typology is enriched to describe the informational aspect of the proposed framework.

4   Proposition and Illustration of a Descriptive Framework

4.1   Proposed Framework

The descriptive framework is composed of three views:

- The <u>physical view</u> of the collective of SPICs: The physical links among SPICs are described using a tree where the nodes correspond to the PI-containers.

Table 1 Scores of existing typologies compared to the informal requirements

Typology		Intelligence level (Req. 2)	Intelligence location (Req. 3)	Aggregation (Req. 4)	Interactions (Req. 5)
Individual mono criterion	Le Moigne [6]	+			
	Wong et al. [7]	+			
	Bajic [8]	+			
	Kiritsis [9]	+			
	Musa et al. [10]	+			
Individual multi-criteria	Meyer et al. [11]	+	+	+	
	Kawsar [12]	+			
	Kortuem et al. [13]	+			
	López et al. [14]	+			
	Sundmaeker et al. [15]	+			
Collective level	Salkham et al. [16]				+
	Iera [17]				+

- The tree is layered according to the three categories of PI-containers (i.e., Transport, Handling and Packaging).
- The links between PI-containers are reported on the tree. A dashed/solid line exhibits respectively the composition/encapsulation mechanism (e.g. Fig. 3).

For each function supported by the collective of SPICs, the informational description is decomposed in two views:

- The "individual intelligence" view: for each SPIC the level of the associated intelligence and its location are specified. The two axes "level of intelligence" and "location of the intelligence" of Meyer's typology are used to support this description.

    - The axis "*level of intelligence*" distinguishes three categories: Information handling, Problem notification and Decision making. These three categories describe respectively entities that can "carry" information, generate alarms relating to their condition and undertake a decision-making process.
    - The axis "*location of intelligence*" is divided into two categories depending on whether the intelligence is embedded in the object or external/remote (i.e. supported via a network).

- The "collective intelligence" view: this view describes the informational links and the interactions among SPICs.

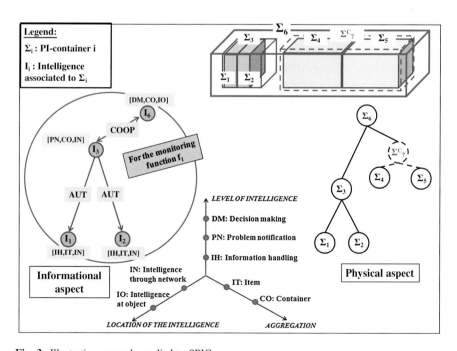

Fig. 3 Illustration example applied to SPIC

– For each SPIC, the third axis "*aggregation*" of Meyer's typology is used to precise if the SPIC can be considered as an intelligent item ("not decomposable" entity) or if the SPIC contains other intelligent items (role of gateway/proxy). (The term "container" in [11] is not related to SPIC case).
– The interactions among the informational systems of the SPICs are described via three relationships:

NUL (*Non-existent*): there is no interaction between the SPICs.
COOP (*Cooperation*): there are interactions between informational systems, but no authority link exists between them. For example, SPICs interact to exchange information on their respective contexts.
AUT (*Authority*): informational systems interact in an authority relationship.

4.2   Example of Applying the Framework

The collective of SPICs considered is the same as the one in Fig. 2. To illustrate the framework, a function f_1 (cargo "monitoring") is considered exploiting the multi-layered intelligence of the collective of SPICs:

- $\Sigma 1$ and $\Sigma 2$ are not equipped with sensors and their status is "monitored" by $\Sigma 3$;
- $\Sigma 4$ and $\Sigma 5$ are assumed containing no perishable goods and are not involved by the monitoring function considered;
- $\Sigma 3$ sends warnings to $\Sigma 6$. This last has decisional capabilities to treat the warnings and to find adequate answers in cooperation with PI management.

Figure 3 depicts the descriptive framework applied in this example. The tree on the right part describes the physical aspect. Concerning the informational aspect (on the left part), the AUT relationships depict that $\Sigma 1$ and $\Sigma 2$ are dependent of $\Sigma 3$. The relationships among the other SPICs are of COOP type. Indeed, $\Sigma 3$, $\Sigma 4$, $\Sigma 5$ and $\Sigma 6$ cooperate to monitor the different cargos. In Fig. 3, the labels associated to the different informational systems are relative to the three axes of Meyer's typology.

5   Conclusion

This paper proposed a framework describing a collective of "smart" PI-containers in interaction. The framework allows the description of the physical links among PI-containers and focuses on the analysis of the informational aspect for a given function. For the informational aspect, two points of view (i.e. individual and collective) are introduced describing the different facets of the interaction situation.

The proposed framework should be considered as a first attempt; it must be validated on several functions implying different intelligence levels and different types of interactions among SPICs.

A short-term prospect aims to improve our analysis framework by refining the facet "interactions". The relationship "cooperation" will be more particularly detailed. Another important prospect is to complement the framework with a methodological guideline allowing choosing the adequate instrumentation according to the functions that must be supported by the multi-layered collective of SPICs.

Acknowledgments The authors like to especially thank the French National Research Agency (ANR) which supports this work via the granted PI-NUTS Project (ANR-14-CE27-0015).

References

1. Montreuil, B.: Towards a physical internet: meeting the global logistics sustainability grand challenge. Logistics Res. **3**(2–3), 71–87 (2011)
2. Montreuil, B., Ballot, E., Tremblay, W.: Modular structural design of physical internet containers. Prog. Mater. Handling Res. **13** (2015)
3. MODULUSHCA (2015). http://www.modulushca.eu/
4. Sallez, Y., Montreuil, B., Ballot, E.: On the activeness of physical internet containers. In: Service Orientation in Holonic and Multi-agent Manufacturing, Springer Studies in Computational Intelligence, pp. 259–269 (2015)
5. Sallez, Y.: Proposition of an analysis framework to describe the "activeness" of a product during its life cycle. In: Service Orientation in Holonic and Multi-Agent Manufacturing and Robotics, Springer Studies in Computational Intelligence, pp. 257–270 (2014)
6. Le Moigne, J.-L.: La théorie du système général: théorie de la modélisation, jeanlouis le moigne-ae mcx (1994)
7. Wong, C.Y., et al.: The intelligent product driven supply chain. In: 2002 IEEE International Conference on Systems, Man and Cybernetics. IEEE (2002)
8. Bajic, E.: Ambient Networking for Intelligent Objects Management, Mobility and Services. Seminar Institute for Manufacturing, IFM Cambridge University, UK (2004)
9. Kiritsis, D.: Closed-loop PLM for intelligent products in the era of the Internet of things. Comput. Aided Des. **43**(5), 479–501 (2011)
10. Musa, A., et al.: Embedded devices for supply chain applications: towards hardware integration of disparate technologies. Expert Syst. Appl. **41**(1), 137–155 (2014)
11. Meyer, G.G., Främling, K., Holmström, J.: Intelligent products: A survey. Comput. Ind. **60**(3), 137–148 (2009)
12. Kawsar, F.: A document-based framework for user centric smart object systems. Ph.D. Thesis, Waseda University, Japan (2009)
13. Kortuem, G., et al.: Smart objects as building blocks for the internet of things. Internet Comput. IEEE **14**(1), 44–51 (2010)
14. López, T.S., et al.: Taxonomy, technology and applications of smart objects. Inf. Syst. Front. **13**(2), 281–300 (2011)
15. Sundmaeker, H., et al.: Vision and challenges for realising the internet of things
16. Salkham, A., et al.: A taxonomy of collaborative context-aware systems. In: UMICS'06, Citeseer (2006)
17. Iera, A.: The social internet of things: from objects that communicate to objects that socialize in the internet. In: Proceedings of 50th FITCE International Congress. Palermo, Italy, Aug 2011
18. Fiske, A.P.: The four elementary forms of sociality: framework for a unified theory of social relations. Psychol. Rev. **99**(4), 689 (1992)

On the Usage of Wireless Sensor Networks to Facilitate Composition/Decomposition of Physical Internet Containers

Nicolas Krommenacker, Patrick Charpentier, Thierry Berger
and Yves Sallez

Abstract Expected to replace current logistic systems, the Physical Internet (PI) motivates to redesign its suitable logistic facilities in which the information system to manage all operations is also changed. In this context, the PI-containers and their encapsulation process (composition/decomposition) are key elements for an open global logistic infrastructure. Although the constitution of a composite container is assumed perfectly known when it was set up, the large variety of manual or automated handling, storage and routing operations in Physical Internet can introduce de synchronization between the real composition of a PI-container and the management information system. The reliance on active objects is a way to overcome this problem. Current RFID technology can serve to identify each container within radio range, but cannot pinpoint its exact location. Fine-grained RFID localization has recently received much attention but these approaches are mainly based on received signals' parameters such as the signal strength (RSSI), time of flight (ToA/TDoA) or angle of arrival. The scope of this paper is to exploit the power of combining spontaneous networking offered by Wireless Sensor Networks with virtual representation. The proposed approach doesn't depend on the quality of received signals that is important in harsh environment. The authors demonstrate that an instantaneous consolidated view of a composite container can be obtained and serve to synchronize both physical and informational flows providing new value-added services.

Keywords Physical internet · Traceability · Wireless sensor networks

N. Krommenacker (✉) · P. Charpentier
Campus Sciences, Université de Lorraine, CRAN CNRS UMR 7039, BP 70239, 54506
Vandœuvre-lès-Nancy Cedex, France
e-mail: nicolas.krommenacker@univ-lorraine.fr

T. Berger · Y. Sallez
Université Lille Nord de France, 59000 Lille, France

T. Berger · Y. Sallez
UVHC, LAMIH UMR n°8201, 59313, Valenciennes, France

© Springer International Publishing Switzerland 2016
T. Borangiu et al. (eds.), *Service Orientation in Holonic and Multi-Agent Manufacturing*, Studies in Computational Intelligence 640,
DOI 10.1007/978-3-319-30337-6_8

1 Introduction

The core of the Physical Internet concept, initiated by Benoit Montreuil, is the handling of standardized modular containers (PI-containers) throughout an open global logistic infrastructure, including key facilities such as PI-Hubs. PI-containers will be manipulated over time (transport, store, load/unload, build/dismantle …) but also, subparts of the containers will be changed (partial loading/unloading, containers splitting and merging). In this context, a significant challenge is to maintain traceability in a highly dynamic transport and logistics system.

The ability to identify the past or current location of an item, as well as to know an item's history, is more complex in Physical Internet due to the wide variety of manual or automated handling, storage and routing operations. In addition to obtain a permanent inventory (full list of delivery items) and the precise location of goods inside all the PI-containers, the traceability system of carried PI-containers and "composite" PI-containers can provide new value-added services:

- Monitoring the conditions throughout the container handling (with sensors deployed to measure temperature, hygrometry …);
- Detect problems for security purposes (e.g., shocks, opening tentative, and incompatibility between goods);
- Guidance information for loading/unloading systems.

Implementing a traceability system requires to systematically link the physical flow of materials and products with the flow of information about them. To avoid synchronization problems between both physical and informational views, we propose to use the Wireless Sensor Networks (WSNs) which are spontaneous multi-hop networks and well-suited for dynamic environments like the Physical Internet. In our approach, each composite PI-container is able to identify its real composition from information collected, and the virtualization of physical PI-containers is used as digital representation of their actual state. The model of the composite container can be updated continuously and is consistent with reality. Hence, the PI-containers play an active role in the PI management and operations [1–4]. Moreover, historical and future states can also be obtained from the virtual representation, and more complex information (unobservable by the human) can be collected and represented [5].

The paper is organized as follows. The PI-containers concept and their composition/decomposition issues in Physical Internet context are introduced in Sect. 2. Section 3 describes the proposed approach based on wireless sensor networks and virtualization to facilitate the traceability of PI-containers. As a proof-of-concept, a composition/decomposition benchmark is used to illustrate the approach and obtained results. Finally conclusive remarks are offered in the last section.

2 PI-Containers Characteristics and Grouping Issues

The following sections offer an overview of the PI-containers concept and of the composition/decomposition process, and assess the situation on current researches.

2.1 Overview of the PI-Container Concept

One of the key concepts of the PI relies on the use of standardized containers that are the fundamental unit loads. Physical goods are not directly manipulated by the PI but are encapsulated in standardized containers, called PI-containers. These containers are moved, handled and stored in the PI network through the different PI-facilities. The ubiquitous usage of PI-containers will make it possible for any company to handle and store any company's products because they will not be handling and storing products per se. More details about key functional specifications of PI-containers can be found in [6, 7].

As introduced in [8], three PI-containers categories can be distinguished: transport, handling and packaging containers. According the Russian doll concept, the three categories can be successively encapsulated one within the other. Figure 1 gives the main characteristics of these categories and their relationships. The modularity enables the containers to better complement each other and therefore allows a better use of the means of transportation.

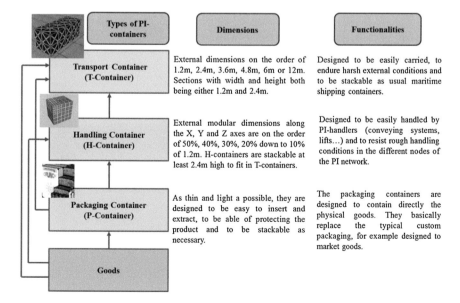

Fig. 1 Relationships between the three categories of PI-containers

As depicted in Fig. 2, the PI-containers can be composed and interlocked to build "composite" PI-containers and allow easier handling or transport, sharing the same standard type of interfacing devices.

2.2   Composition/Decomposition Issues in Physical Internet Context

The PI relies on a distributed multi-segment intermodal network. By analogy with the Digital Internet that transmits data packets rather than information/files, the PI-containers constitute the material flow among the different nodes of the PI network. The design of cross-docking hub (in analogy with digital internet, can be seen as a router), allowing the quick, flexible and synchronized transfer of the PI-containers, is essential for the successful development of the Physical Internet. Different types of hubs, denoted PI-hubs, are considered (e.g., road to rail, road to road, ship to rail).

The efficient management of PI-hubs is a cornerstone for the PI development and the composition/decomposition process is a key process to master. Figure 3 illustrates the treatment of composite H-containers contained in T-containers.

Several steps can be distinguished:

1. Unloading: At their arrival in the PI-hub, the T-containers are unloaded from their inbound carriers. They are either standalone or composite.
2. Preparation (Composition/Decomposition): According to the appropriate degree of decomposition, the cargo of the T-containers are decomposed to obtain standalone H-containers or groups of H-containers. The groups correspond to several H-containers kept together because they have the same next destination

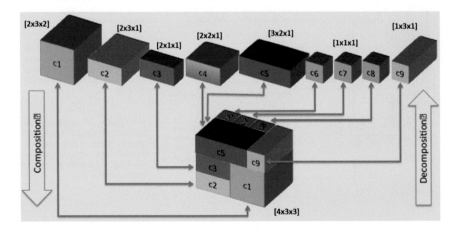

Fig. 2 Example of composition/decomposition of PI-containers (given in [7])

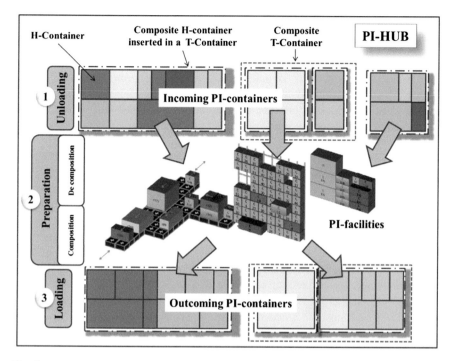

Fig. 3 Example of PI-containers treatment in a PI-hub

in the PI network. The resulting H-containers are composed adequately with other H-containers to constitute the cargos for outbound T-containers. The challenge is to create composed sets of H-containers that fit in the available spaces within the assigned outbound T-container. In this last, the sets of H-containers can be separated into several groups according to their unloading destinations.

3. Loading: In timely fashion, the composed cargos are inserted in the T-containers and moved to their assigned docks to allow loading in the outbound carriers.

At each step of this process, the constitution of each composite container is assumed perfectly known. In case of problems occurring at the composite container (e.g. constituent lost or stolen), the correspondence between the real composite container and its model is false.

This information loss can imply a longer time for the PI-hub management system to reorder in the correct way. Any mistake in the composition/decomposition process can cause negative impacts on the overall PI-hub performance in terms of delays and cost of operations. The following section gives an overview of the current research works in the field of composition/decomposition in PI-hubs.

2.3 Current Researches on Composition/Decomposition of PI-Containers

The design of PI-containers is under study in the European MODULUSHCA project [9] aiming to build and test physical prototypes of H-containers. The mechanisms of interlocking between containers are more particularly studied. This functionality is mandatory to allow quick and efficient composition/decomposition processes.

At this point of development of the Physical Internet, very few research works have addressed the decomposition/composition problems in a PI-hub. In [10], the flow of PI-containers in the PI network is studied in simulation according a macroscopic view. How the composition/decomposition process is physically realized is not studied in detail up to now.

In [11, 12], the authors study by simulation the handling of PI-containers in different types of PI-hubs. The decomposition/composition problem is partially addressed because only PI-containers of same type (T-containers) are considered. The PI-containers are juxtaposed according to a linear pattern and more complex compositions, such as those illustrated in Fig. 2, are not taken into account. The composition/decomposition process is treated in a centralized way. Each composite container is identified (via by example a RFID tag) and the list of its constituents is accessible through network.

In [13] the authors propose a distributed approach and realize the grouping of T-containers using interactions between "smart" PI-containers but, as in the previous works, all the PI-containers are of the same type.

In all these works, in the case of problem arising on the composite container as stated in Sect. 2.2, the model no longer corresponds to the reality. To face this significant limitation, a distributed approach able to identify the real composition of a composite PI-container is presented in the next section of the paper. The originality of this approach is that the results don't depend on the quality of the wireless signals transmitted.

3 The Proposed Approach to Control Composite PI-Containers

To deal with the composition/decomposition issues in PI-hubs, we propose a Virtualization of Container (VoC) framework that consolidates assignment information of PI-containers in a composite container. This virtual environment allows PI-operators to visualize anywhere and at any time, the real composition of a composite PI-container.

For that purpose, we use a wireless sensor network (WSN) where nodes are attached to each container, and store information about the container such as the container category, the identifier and its dimensions. The sensor node embedded at the composite container level, acts as a gateway and provides the interface between the management information system (or PI-operators) and the composite container. According to the transmission range, a spontaneous multi-hop network is formed. Through their cooperation and the execution of a neighbour discovery protocol, the one-hop neighbour table is computed.

A Constraint Satisfaction Problem (CSP) can be formulated where:

- The neighbour table gives constraints related to positions between the unitary-containers (allocation restrictions);
- The container dimensions provide basic geometric constraints. The unitary-containers lie entirely within the composite container and do not overlap. Each one of them may only be placed with its edges parallel to the walls of the composite container.

Therefore, each feasible solution of the CSP is a potential loading pattern and the 3D container virtualization process provides an instantaneous consolidated view (dynamic and virtual) of the composite container assemblage, as depicted in Fig. 4. The mathematical formulation of this satisfaction problem is similar to the well-known 3D Container Loading Problem with a single container and a number of heterogeneous boxes [14].

However, the objective is not to optimize the number of items which have to be packed, but to find the assignment that satisfies all constraints and matches with the real composition of the H-container.

This issue depends directly on the transmission range and the dimensions of the containers. The number of feasible solutions is indeed related to the neighbour graph connectivity, obtained from the neighbour table. Assuming the same transmission range for all nodes, two nodes are neighbours if they can communicate, i.e. the distance between them is less than or equal to the transmission range. Therefore,

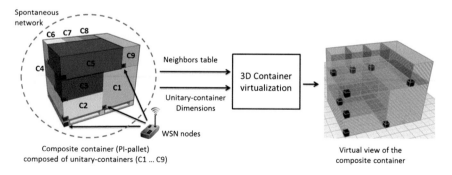

Fig. 4 Virtualization of Container (VoC) framework

with a transmission range smaller than the smallest container, a lot of nodes will be unable to communicate.

The set of allocation constraints in the CSP will be reduced, leading to many feasible solutions. Similarly, if each node can communicate with all the other, the neighbour graph will be a complete graph. In this case, multiple feasible solutions can be found from a simple permutation of two containers with the same dimensions. The transmission range plays an important role to limit the number of feasible solutions and obtain the virtual view of the composite container.

The set of variables, constraints and the mathematical formulation of the CSP can be found in [15]. As a proof-of-concept, a composition/decomposition scenario is used to illustrate the approach. The simulation scenario and results are presented in the next section.

4   Simulations and Results

The composition and decomposition scenario presented in [9] is here used. A composite H-container is constituted with 9 unitary P-containers. Figure 2 illustrates this scenario with the normalized container dimensions. The assortment is strongly heterogeneous with 6 categories of container's dimensions (C5, C2 and C6, C7, C9 are similar).

As shown Fig. 4, the unitary-containers are placed so that they fit perfectly the volume of the composite PI-container, here a PI-pallet.

Although not necessary, each embedded sensor node is at the front-left-bottom corner for all containers. This situation can occur when interlocking mechanisms limit the pivotal function. The sensor node fixed to the PI-pallet acts as the gateway and collects all neighbourhood and dimension information. Several data sets are considered with a transmission range between 1 and 3 (from the smallest to the highest container dimension respectively).

The CSP is implemented in the Matlab optimization toolbox and results are obtained with the *fmincon* solver.

The neighbour graph and the number of feasible solution S for different transmission range R are illustrated in Fig. 5. When the neighbour graph is almost maximally ($R = 3$) or minimally ($R = 1$) connected, the number of solutions is large and the real packing pattern cannot be determine. On the other hand, a value ($R = 2$) gives only one feasible solution with a complying packing pattern. So this solution provides the real composition of the composite PI-container.

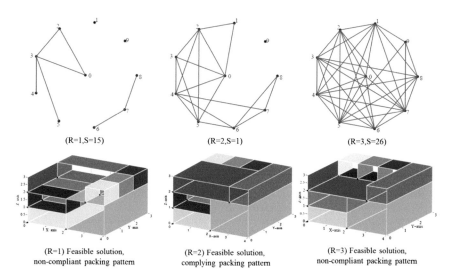

(R=1,S=15)   (R=2,S=1)   (R=3,S=26)

(R=1) Feasible solution,
non-compliant packing pattern

(R=2) Feasible solution,
complying packing pattern

(R=3) Feasible solution,
non-compliant packing pattern

Fig. 5 CSP solutions and container packing patterns

5 Conclusion

The large variety of manual or automated handling, storage and routing operations characterizes the Physical Internet as a highly dynamic transport and logistics system. In this paper, we have focused on the traceability of containers in Physical Internet context in which the management information system must be redesigned.

To avoid any mistake in the composition/decomposition process, the real composition of a composite PI-container must be known at all times. To do this, we have proposed an approach based on Wireless Sensor Networks (WSN) and a VoC framework to consolidate assignment information of PI-containers in the composite container.

A simulation demonstrated that the real 3D pattern can be obtained from the cooperation between nodes. Our approach, although more expensive than RFID technology, offers the benefit from knowing the exact location (through the virtual representation). The WSN technology could also serve to support information about containers, or generate new information based on sensing capabilities.

References

1. Sallez, Y., Montreuil, B., Ballot, E.: On the activeness of physical internet containers. In: Borangiu, T., Trentesaux, D., Thomas, A. (eds.) Service Orientation in Holonic and Multi-agent Manufacturing. Springer series Studies in Computational Intelligence, Vol. 594, pp. 259–269 (2015)

2. Wong, C.Y., McFarlane, D., Zaharudin, A.A., Agarwal, V.: The intelligent product driven supply chain. In: IEEE International Conference on Systems, Man and Cybernetics, Hammamet, Tunis (2002)
3. Sallez, Y.: Proposition of an analysis framework to describe the "activeness" of a product during its life cycle—part I: Method and applications. In: Borangiu, T., Trentesaux, D. (eds) Service Orientation in Holonic and Multi-Agent Manufacturing Control, Studies in Computational Intelligence, vol. 544, pp. 271–282. Springer (2014)
4. Sallez, Y.: The augmentation concept: how to make a product "active" during its life cycle. In: Borangiu, T., Trentesaux, D. (eds) Service Orientation in Holonic and Multi-Agent Manufacturing Control, Studies in Computational Intelligence, vol. 402, pp. 35–48. Springer (2012)
5. Verdouw, C.N., Beulens, A.J.M., Reijers, H.A.: A control model for object virtualization in supply chain management, Computers in Industry, Vol. 68, 116–131, Apr 2015
6. Ballot, E., Montreuil, B., Meller, R.D.: The Physical Internet: The Network of the Logistics Networks. La Documentation Française, Paris (2014)
7. Montreuil, B.: Towards a physical internet: meeting the global logistics sustainability grand challenge. Logistics Res. 3(2–3), 71–87 (2011)
8. Montreuil, B., Ballot, E., Tremblay, W.: Modular Structural Design of Physical Internet Containers. Prog. Mater. Handling Res. 13 (2014) (MHI)
9. Modulushca project (2015). http://www.modulushca.eu/
10. Ballot, E., Montreuil, B., Thémans, M.: OPENFRET: contribution à la conceptualisation et à la réalisation d'un hub rail-route de l'Internet Physique. MEDDAT, Paris (2010)
11. Ballot, E., Montreuil, B., Thivierge, C.: Functional Design of Physical Internet Facilities: A Road-Rail Hub. Progress in Material Handling Research, MHIA, Charlotte, NC (2012)
12. Meller, R.D., Montreuil, B., Thivierge, C., Montreuil, B.: Functional Design of Physical Internet Facilities: A Road-Based Transit Center. Progress in Material Handling Research, MHIA, Charlotte, NC (2012)
13. Pach, C., Sallez, Y., Berger, T., Bonte, T., Trentesaux, D., Montreuil, B.: Routing management in physical internet cross docking hubs: study of grouping strategies for truck loading. In: International Conference on Advances in Production Management Systems APMS, IFIP AICT, vol. 438, pp. 483–490. Springer, Sept 2014
14. Bortfeldt, A., Wäscher, G.: Container loading problems—a state-of-the-art review. Otto-von-Guericke Universität Magdeburg, Working Paper No. 7/2012 (2012)
15. Tran-Dang, H., Krommenacker, N., Charpentier, P.: Enhancing the functionality of physical internet containers by WSN. In: International Physical Internet Conference. Paris, July 2015

Part III
Sustainability Issues in Intelligent Manufacturing Systems

Part II
Sustainability Issues in Intelligent
Transportation Systems

Artefacts and Guidelines for Designing Sustainable Manufacturing Systems

Adriana Giret and Damien Trentesaux

Abstract The following key questions are the main focus of this paper: Which are the needs to integrate sustainability and efficiency performances in Intelligent Manufacturing System design? And: How can these needs be approached using concepts from Intelligent Manufacturing System engineering methods in the context of design of sustainable manufacturing systems? This paper answers these questions with: "green" artefacts and guidelines for helping to maximize production efficiency and balance environmental constraints already in the system design phase. In this way the engineers designing the manufacturing system can have guidelines for decision support and tools for improving energy efficiency, CO_2 emissions and other environmental impacts integrated into a software engineering method for intelligent manufacturing development.

Keywords Sustainable manufacturing systems · Multi-agent system · Holonic manufacturing system · Intelligent manufacturing design

1 Introduction

There is now a well-recognized need for achieving overall sustainability in manufacturing activities [1–4], due to several established and emerging causes: environmental concerns, diminishing non-renewable resources, stricter legislation and inflated energy costs, increasing consumer preference for environmentally friendly products, etc. In order to achieve sustainability in production all the components,

A. Giret (✉)
Dpto. Sistemas Informaticos Y Computacion, Universidad
Politecnica de Valencia, Valencia, Spain
e-mail: agiret@dsic.upv.es

D. Trentesaux
LAMIH UMR CNRS 8201, University of Valenciennes
and Hainaut-Cambrésis, Valenciennes 59313, France
e-mail: Damien.Trentesaux@univ-valenciennes.fr

© Springer International Publishing Switzerland 2016
T. Borangiu et al. (eds.), *Service Orientation in Holonic and Multi-Agent Manufacturing*, Studies in Computational Intelligence 640,
DOI 10.1007/978-3-319-30337-6_9

93

processes and performance indicators must be taken into account at all relevant levels (product, process, and system). One of the key questions to answer in the field of Sustainable Production is: What approaches should/could be used to transform production processes to be more sustainable? The authors believe that to foster sustainability in production the whole lifecycle of manufacturing systems must be taken into account, considering its different layers in a holistic way. From systems' conception throughout implementation, until maintenance the system developer must take into account sustainability issues. Nevertheless, there is a lack of sustainability considerations in the state-of-the-art design methods for manufacturing operations [5–7]. Despite that other relevant levels have a large number of approaches that take special consideration to sustainability issues (for a state-of-the-art review see for example [1, 8, 9]). To fill the gap, in this paper a design artefact and a set of guidelines for the development of sustainable manufacturing systems are proposed.

2 Sustainable Manufacturing and Its Design

Salonitis and Ball presented in [10] the new challenges imposed by adding sustainability as a new driver in manufacturing modelling and simulation. This very complex and challenging undertake must also consider issues at all relevant levels in manufacturing—product, process, and system [11].

It is crucial and urgent for system engineers of sustainable manufacturing systems to have tools and methods that can help them to undertake this task from system conception, trough out its design until its execution in an effective way. The research field of Intelligent Manufacturing Systems (IMS) provides a large list of engineering methods tailored to deal with specific aspects for designing IMS (for a comparative study see [5]). Nevertheless, most of the existing approaches do not integrate specific support for designing sustainable manufacturing systems. One of the major challenges in developing such approaches is the lack on guidelines and tools that foster the system designer to consider sustainability issues at design phases and that can help during the implementation of the IMS. Then two key questions to answer are the following: (Q1) What are the needs to integrate sustainability efficiency performance in IMS design? [12, 13] (Q2) How can these needs be approached using concepts from IMS engineering methods in the context of designing sustainable manufacturing systems?

The authors believe that integrating sustainability efficiency performance in IMS design can be tackled by means of:

- Specific guidelines that can help the system designer to know (1) what sustainability parameters are key to the system, (2) how these parameters must be taken into account by the components of the IMS, (3) when these parameters must be used for achieving sustainable efficiency in the system, (4) which approaches can be used to compute a sustainable solution for the different tasks and processes of the manufacturing system.

- "Green" artefacts that can provide optimized solutions for concrete aspects at different levels such as: enterprise resource planning, production control, manufacturing operations scheduling, etc.

The above mentioned aspects, which are some answers for Q1, are the main focus of this paper. Moreover, this paper answers Q2 by means of a specific approach for IMS development called Go-green ANEMONA [14]. The authors believe that the answers provided in this paper for Q1 are two of a larger list. Finding out which are the complete elements of this list is outside the paper's scope and an open problem worth for a deeper study. In this paper the complete details of sustainable specific guidelines and green artefacts to assist the system engineer during the IMS design are described. Moreover, the engineering process is showcased with a case study.

3 Go-Green ANEMONA

Go-green ANEMONA [14] provides the methodological benefits of holons and multi-agent systems for the identification and specification of specific sustainability features of manufacturing systems. It is the integration of a previously released and already sound Multi-agent engineering method for IMS [15], and the new go-green holon concept [16] that foster system designers to bring sustainability features into manufacturing operations control architectures. They are focused on the identification of manufacturing holons and the design and integration of sustainability-oriented mechanisms in the system specification. Figure 1 shows the development process of Go-green ANEMONA. In this figure it can be noticed that the process is an iterative and recursive sequence of specific activities to specify, analyse, design, implement, deploy and maintain the IMS. Every step in the process is augmented with a set of specific guidelines for concrete aspects during system development [14]. At the same time, at each step during analysis, design, implementation, deployment and maintenance a specific artefact—the Go-green Holon is provided, which the system engineer can use in order to design and implement sustainability problem solving methods.

In the following sections the details of the Go-green Holon as a design artefact and the sustainability guidelines for designing IMS are described.

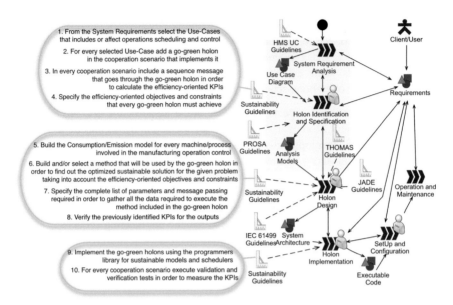

Fig. 1 Go-green ANEMONA process

3.1 Go-Green Manufacturing Holons: An Artefact for Designing and Developing Sustainable IMS

A **Go-green manufacturing Holon** [16] is a holon that, in the context of sustainable manufacturing, considers complementary efficiency-oriented mechanisms, in addition to classical effectiveness-oriented mechanisms, to make a decision and/or execute an operation. Go-green manufacturing holons may apply different solving approaches: a balanced compromise (between effectiveness and efficiency), a lexicographical-oriented decision making process (e.g., optimize first effectiveness, then efficiency in an opportunistic way) or a constrained problem (e.g., optimize efficiency under effectiveness constraints). Multicriteria analysis, simulation and operations research approaches can be useful in this context.

In a holarchy, classical and go-green manufacturing holons may coexist. Also, go-green manufacturing holons may address only efficiency issues and could for example, cooperate with classical holons to reach a global consensus. Figure 2 shows an example of a Go-green (manufacturing) resource Holon that integrates concrete capabilities for dealing with sustainability efficiency performance in a resource. In this way when the system engineer needs redesigning a given manufacturing resource for tackling for example energy-efficiency in its operations, he/she can use the Go-green resource Holon that is a pre-built artefact with built-in functionalities that can be parameterized and or fine-tuned in order to design the concrete resource with its concrete energy-efficiency parameters. The Go-green resource Holon can seamlessly interact with "classical" holons (other resource

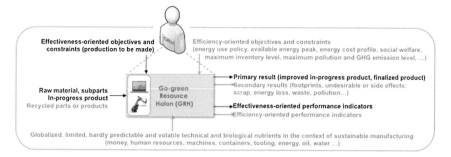

Fig. 2 A Go-green Resource Holon

holons, and/or product holons, work-order holons and staff holons) since the Go-green ANEMONA metamodel provides the support for implementing the cooperation with them.

3.2 Sustainability Guidelines

The *Sustainability Guidelines* (Fig. 1) are built-in specification guidelines of Go-green ANEMONA to assist the system engineer in: (i) finding out in which cooperation scenarios a Go-green Holon is required; (ii) what will be the interaction protocol (the complete message sequence) among Go-green Holons and classical holons; (iii) which parameters must be specified in order to complete the specification of a Go-green Holon; (iv) what models must be defined in order to complete their implementation; etc. Go-green ANEMONA provides support for implementing Go-green Holons by means of a library of pre-built solving approaches from which the engineer can select the type of service which better suits his/her needs for efficiency-oriented objectives, constraints and KPIs (Key Performance Indicators).

It is important to point out that guideline 6 is supported by a decision flow diagram [8] that enforces researchers to explicit and to think about their main designs choices (a design choice being the choice of the best categories for each of the sustainable parameters taken into account in the IMS) when designing a specific sustainable Go-green Holon. Let's imagine, for example, a situation in which a Go-green Holon is in charge of production scheduling, and the solution must take into account energy use, the number of machines, and CO_2 emissions. At the same time, the scheduling must be achieved off-line, but must be adapted to react to run-time events such as machine breakdowns, new work orders entering the system, and variations in energy consumption. In addition, there are thresholds for makespan, energy use (peak power consumption, etc.), and CO_2 emissions (quota). Finally, the makespan must be optimized, energy economized, and CO_2 emissions reduced. The two last requirements will determine the way the multi-objective must be handled. For this concrete situation, a solving approach is required that: takes into account energy and CO_2; maintains the scheduling effectiveness as the main objective while minimizing energy and CO_2, and; is a proactive-reactive scheduling method (an initial schedule is computed off-line and re-scheduling activities are executed on-line). With this decision support the system engineer can chose from the library of pre-built solving approaches the one that better fits these requirements (see [8] for a list of approaches suitable for different sustainable requirements combinations).

4 Case Study

For illustration and proof of feasibility, an intelligent distributed monitoring and control application of a ceramic tile factory is designed using Go-green ANEMONA. The sustainable goals to optimize in this application are: minimize scraps and waste of materials due to bad quality of the tiles, and minimize the energy consumption by re-using the oven residual heat in the drying stage of the production line. Apart from these goals the IMS must also minimize the makespan.

Figure 3 shows a diagram in which a *Scraps and Waste Go-green Holon* is designed with the goals: "minimize scraps and waste in the tile press machine", "assure a correct cooking of the tile", "re-direct tile scraps to the clay mix", "find-out the better production sequence of tiles' work orders to minimize scraps due to press configuration changes". From guideline 6 the *Scraps and Waste Go-green Holon* takes an approach for scraps and waste minimization using a greedy randomized adaptive search [17] in order to find out the optimized sequence of work orders for minimizing scraps due to press configuration change. This approach is used in the *Adjust Resource* task in Fig. 3.

The case study developed with Go-green ANEMONA was compared with a previous development in which the system engineer designed the system without the specific guidelines or the Go-green Holons. Table 1 shows the results.

From Table 1 it can be noticed that the number of iterations to find out the set of holons that implements the IMS is the same, but in terms of duration Go-green ANEMONA outperforms ANEMONA by 0.5 months (it is important to point out that the developers' team for both developments were different but with the same number of members and the same skills on IMS design).

To measure the *Easy to Design* and the *Guidelines Usefulness* MASEV, a MAS evaluation framework was used [18]. This framework allows analysing and comparing methods and tools for developing MAS in terms of general requirements and method guidelines. There is a questionnaire in which system engineers answer different questions related to the aspect that is being evaluated; a numerical value is assigned to each answer. It can be noticed that Go-green ANEMONA got 9.5 out of 10 when evaluating the usefulness of Go-green Holons (Easy to Design), and 9.2 out of 10 when evaluating the usefulness of specific guidelines to design sustainable IMS.

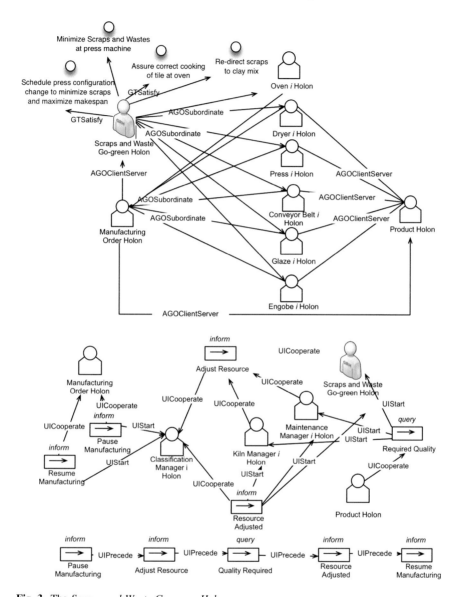

Fig. 3 The *Scraps and Waste Go-green Holon*

On the other hand, when evaluating the no. of Holons identified with Go-green ANEMONA it turns out that 8 more holons were identified compared with the ANEMONA development. This is because the Go-green Holons are added to the classical holons in the development. But the Go-green Holons helped to have less cooperation domains with Go-green ANEMONA since there is no need to have

Table 1 Case study comparison: ANEMONA versus Go-green ANENOMA

Method	Time to design	Easy to design	Guidelines usefulness	No. of holons	No. of Co. Do.
ANEMONA	3 iterations (3 months)	7.5/10	7/10	46	34
Go-green ANEMONA	3 iterations (2.5 months)	9.5/10	9.2/10	54	28

such cooperation domains for sustainability issues because they are already taken into account in the different cooperation domains in which the Go-green Holons are involved.

5 Conclusions

In this paper the answers to the following questions where analysed: (Q1) What are the needs to integrate sustainability efficiency performance in IMS design? (Q2) How can these needs be approached using concepts from IMS engineering methods in the context of sustainable manufacturing systems design? The main proposals for answering the questions are: (1) a Go-green holon, as a green artefact that helps the system designer to implement solutions for sustainable IMS, and (2) a set of guidelines that enforces system engineers to think about their main designs choices of the sustainable parameters taken into account in the IMS. The proposal was showcased designing an intelligent distributed monitoring and control application of a ceramic tile factory. Nevertheless, the guidelines and artefacts have helped in the development of the case study; the authors believe that these are only 2 of a larger list of design elements for developing sustainable manufacturing systems; this global list is still open to study.

The proposed approach is still under development. The library of pre-built solving methods from which the engineer can select the type of service which better suits his/her needs for the efficiency-oriented objectives, constraints and KPIs of go-green holons is being populated. Moreover, a case-tool is being designed as design support.

References

1. Garetti, M., Taisch, M.: Sustainable manufacturing: trends and research challenges. Prod. Plan. Control **23**, 83–104 (2012). doi:10.1080/09537287.2011.591619
2. Fang, K., Uhan, N., Zhao, F., Sutherland, J.W.: A new approach to scheduling in manufacturing for power consumption and carbon footprint reduction. J. Manuf. Syst. **30**, 234–240 (2011)

3. Merkert, L., Harjunkoski, I., Isaksson, A., Säynevirta, S., Saarela, A., Sand, G.: Scheduling and energy-industrial challenges and opportunities. Comput. Chem. Eng. **72**, 183–198 (2015)
4. Evans, S. Bergendahl, M., Gregory, M., Ryan, C.: Towards a sustainable industrial system. with recommendations for education, research, industry and policy. http://www.ifm.eng.cam. ac.uk/uploads/Resources/Reports/industrial_sustainability_report (2009)
5. Giret, A., Trentesaux, D.: Software engineering methods for intelligent manufacturing systems: a comparative survey. Ind. Appl. Holonic Multi-Agent Syst. 11–21 (2015)
6. Thomas, A., Trentesaux, D.: Are intelligent manufacturing systems sustainable? In: Borangiu. T., Trentesaux, D., Thomas, A., (ed.) Service Orientation in Holonic and Multi-Agent Manufacturing and Robotics, Springer Studies in Comput. Intell., pp. 3–14
7. Matsuda, M., Kimura, F.: Usage of a digital eco-factory for green production preparation. Procedia CIRP. **7**, 181-186. ISSN 2212-8271 (2013)
8. Giret, A., Trentesaux, D., Prabhu, V.: Sustainability in manufacturing operations scheduling: a state of the art review. J. Manuf. Syst., To appear (2015)
9. Badurdeen, F., Iyengar, D., Goldsby, T.J., Metta, H., Gupta, S., Jawahir, I.S.: Extending total life-cycle thinking to sustainable supply chain design. Int. J. Prod. Lifecycle Manage. **4**(49), 6 (2009)
10. Salonitis, K., Ball, P.: Energy efficient manufacturing from machine tools to manufacturing systems. Procedia CIRP. 7:634–639, ISSN 2212-8271 (2013)
11. Jayal, A.D., Badurdeen, F., Dillon Jr, O.W., Jawahir, I.S.: Sustainable manufacturing: modeling and optimization challenges at the product, process and system levels. CIRP J. Manuf. Sci. Technol. **2**, 144–152 (2010). doi:10.1016/j.cirpj.2010.03.006
12. Taticchi, P., Tonelli, F., Pasqualino, R.: Performance measurement of sustainable supply chains: a literature review and a research agenda. Int. J. Prod. Perform. Manage. **62**(8), 782–804 (2013)
13. Taticchi, P., Garengo, P., Nudurupati, S.S., Tonelli, F., Pasqualino, R.: A review of decision-support tools and performance measurement and sustainable supply chain management. Int. J. Prod. Res. **53**(21), 6473–6494 (2015)
14. Giret, A., Trentesaux, D.: Go-Green Anemona: a manufacturing system engineering method that fosters sustainability. Glob. Clean Prod. Sustain. Cons. Conf, To appear (2015)
15. Giret, A., Botti, V.: Engineering holonic manufacturing systems. Comput. Ind. **60**, 428–440 (2009). doi:10.1016/j.compind.2009.02.007
16. Trentesaux, D., Giret, A.: Go-green manufacturing holons: a step towards sustainable manufacturing operations control. Manuf. Lett. **5**, 29–33 (2015)
17. Escamilla, J., Salido, M.A., Giret, A., Barber, F.: A Metaheuristic technique for energy-efficiency in job-shop scheduling. Proc. Constraint Satisf. Tech. COPLAS, 24th Int. Conf. Autom. Plan. Sched. ICAPS'14 (2014)
18. Garcia, E., Giret, A., Botti, V.: Evaluating software engineering techniques for developing complex systems with multiagent approaches. Inf. Soft. Technol. **53**, 494–506 (2011)

A Human-Centred Design to Break the Myth of the "Magic Human" in Intelligent Manufacturing Systems

Damien Trentesaux and Patrick Millot

Abstract The techno-centred design approach, currently used in industrial engineering and especially when designing Intelligent Manufacturing Systems (IMS) voluntarily ignores the human operator when the system operates correctly, but supposes the human is endowed with "magic capabilities" to fix difficult situations. But this so-called magic human faces with a lack of elements to make the relevant decisions. This paper claims that the Human Operator's role must be defined at the early design phase of the IMS. We try to show with examples of systems from manufacturing as well as from energy or transportation that the Human Centred Design approaches place explicitly the "human in the loop" of the system to be automated. We first show the limits of techno-centred design methods. Secondly we propose the principles of a balanced function allocation between human and machine and even a real cooperation between them. The approach is based on the system decomposition into an abstraction hierarchy (strategic, tactical, operational). A relevant knowledge of the human capabilities and limits leads to the choice of the adequate Level of Automation (LoA) according to the system situation.

Keywords Techno-centred design · Human centred design · Human in the loop · Levels of automation · Human-machine cooperation · Intelligent manufacturing systems

D. Trentesaux (✉) · P. Millot
LAMIH, UMR CNRS 8201, University of Valenciennes
and Hainaut-Cambrésis, UVHC, Le Mont Houy, Valenciennes, France
e-mail: damien.trentesaux@univ-valenciennes.fr

P. Millot
e-mail: patrick.millot@univ-valenciennes.fr

© Springer International Publishing Switzerland 2016
T. Borangiu et al. (eds.), *Service Orientation in Holonic and Multi-Agent Manufacturing*, Studies in Computational Intelligence 640,
DOI 10.1007/978-3-319-30337-6_10

1 Introduction

This paper is relevant to industrial engineering, energy and services in general, but is focused on Intelligent Manufacturing Systems (IMS). It deals with the way the human operator is considered from a control point of view when designing IMS that integrates human beings.

The complexity of industrial systems and human organizations that control them is increasing with time, as well as their required safety levels. These requirements evolve accordingly with past negative experiences and industrial disasters (Seveso, Bhopal, AZF, Chernobyl…). In France, the Ministry for Ecology, Sustainable Development and Energy (Ministère de l'Écologie, du Développement durable et de l'Énergie) has led a study in technological accidents that occurred in France in 2013 ("inventaire 2014 des accidents technologiques"). It has shown that the three first domains expressed in terms of numbers of accidents are manufacturing, water and waste treatment. This study has also highlighted that even if "only" 11 % of the root causes come from a "counter-productive human intervention", human operators are often involved in accidents at different levels: organizational issues; default in control, monitoring and supervision; bad equipment choice; and lacks in knowledge capitalization from past experiences.

Obviously, the capabilities and limits of the human operator during manufacturing have been widely considered for several years, and very intensively by industrialists. This attention has been mainly paid at an operational level:

- At the physical level: industrial ergonomic studies, norms and methods (MTM, MOST…) are a clear illustration of this;
- At the informational and decisional levels: industrial *lean* and *kaizen* techniques aim to provide the operator with informational and decisional capabilities to react and to improve the manufacturing processes for predefined functioning modes of the manufacturing system.

Meanwhile, these industrialist-oriented technical solutions lack human-oriented considerations when dealing with higher and more global decisional and informational levels such as scheduling, supervision, etc. as well as when abnormal and unforeseen situations and modes occur. This holds also true for the related scientific research activity. And this is even truer for less mature and more recent research topics such as those dealing with the design of control in IMS architectures. In addition, and specifically to IMS, where it is known that emerging (unexpected) control behaviours can occur during manufacturing, the risk to face possible accidents or unexpected and possibly hazardous situations when using un-human-aware control systems increases.

The objective of this paper is then to foster researchers dealing with the design of control systems in IMS to question the way they consider the real capabilities and limitations of the human beings. It is important to note that, at our stage of development, this paper remains highly prospective and contains only a set of human-oriented specifications that we think researchers must be aware of when designing their control in IMS. For that purpose, before providing these specifications, the following part describes the consequence of designing un-human-aware control systems in IMS, which corresponds to what we call a "techno-centred" approach.

2 The Classical Control Design Approach in IMS: A Techno-Centred Approach

As introduced, we consider in this paper the way the human operator is integrated within the control architectures in IMS. Such "Human-in-the-loop" Intelligent Manufacturing Control Systems are denoted, for simplification purpose, HIMCoS in this paper. These systems consider the intervention of human (typically, information providing, decision making or direct action on physical components) during the intelligent control of any functions relevant to the operational level of manufacturing operations, being for example scheduling, maintenance, monitoring, inventory management, supply, etc. Intelligence in manufacturing control refers to the ability to react, learn, adapt, reconfigure, evolve, etc. with time using computational and artificial intelligence technics, the control architecture being typically structured using Cyber-Physical Systems (CPS) and modelled using multi-agent or holonic principles, in a static or dynamic way (i.e., embedding self-organizational capabilities). The intervention of the human is limited in this paper to the decisional and information aspects (we do not consider direct and physical action on the controlled system for example).

2.1 An Illustration of the Techno-Centred Approach in IMS

To illustrate what we call the techno-centred design approach in this context, let us focus and consider a widespread studied IMS domain: distributed scheduling in manufacturing control. Research activities in this domain foster a paradigm that aims to provide more autonomy and adaptation capabilities to the manufacturing control system by distributing functionally or geographically the informational and decisional capabilities among artificial entities (typically agents or holons). This paradigm creates "bottom-up" emerging behavioural mechanisms complementarily to possible "top-down" ones generated by a centralized and predictive system to limit or to force this emerging behaviour evolving within pre-fixed bounds [1]. This paradigm encourages designers to provide these entities with cooperation or negotiation skills so that they can react and adapt more easily to the growing level

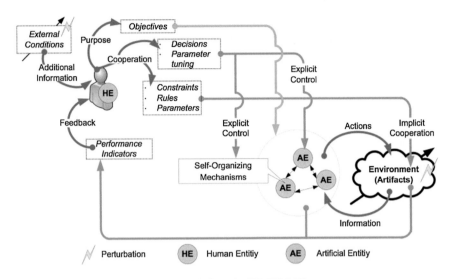

Fig. 1 An example of a techno-centred design of a HIMCOS [8]

of uncertainty in the manufacturing environment while controlling the complexity of the manufacturing system by splitting the global control problem into several local ones.

The most emblematic proposals relevant to this paradigm are PROSA [2], ADACOR [3], and more recently ADACOR^2 [4]. Nowadays, the initial paradigm has evolved but basics remain the same: up-to-date proposals are typically linked to the concepts of "intelligent product" [5] and "cyber-physical systems" [6]. Researchers in this domain (and more generally in industrial engineering) often consider that the human operator is the supervisor of the whole [7, 8]. Figure 1 (from [8]) is a typical illustration. According to this approach, the human operator fixes the objectives, tune parameters, constraints and rules. He then influences or explicitly controls artificial entities. Even if such works try to integrate the human operators, a lot of other ones do not even consider this possibility, which is inconsistent in a real life context.

This is a usual design approach in manufacturing and industrial engineering. Paradoxically, it can also be identified even in the widespread and historical field of Decision Support Systems (DSS), see for example [9, 10] or [11]. This design approach can be characterized as "techno-centred", which means that priority is assigned to the solving of technical issues to the detriment of the human aspects. A techno-centred approach consists in automating a maximum number of functions (in nominal or degraded mode) in a pre-defined context and in assuming that the human operator will supervise and handle all the situations that where not foreseen.

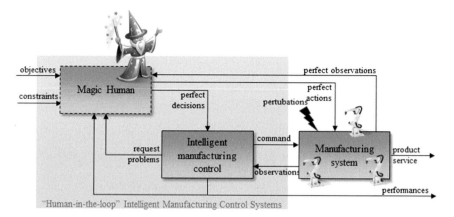

Fig. 2 Techno-centred HIMCoS design approach

2.2 The Hidden Assumption of the "Magic Human"

In fact, in a techno-centred design of HIMCoS, there is a *hidden assumption*: the human operator is considered as an omniscient person that will:

- Solve all the problems for which there is no anticipated solving process,
- Provide the good information to the control system in due time,
- Decide among alternatives in real time (i.e., as fast as possible or in due date), whenever it is necessary (cf. the concept of DSS),
- Ensure with full reliability the switching between control configurations and especially the recovery towards normal operating conditions after unforeseen perturbation/degradation.
  With derision, we call him the *magic human* (Fig. 2).

As a synthesis of our point of view, Fig. 2 sums up the techno-centred design approach in HIMCoS. In this figure, the dotted lines represent the fact that the human operator is not really considered during the design of the HIMCoS.

2.3 The Risks of Keeping This Assumption Hidden

Assuming the human operator a magic human is obviously not realistic but it is a reality in research in industrial engineering. In light of the mentioned reference in the introduction to the French ministry study, a techno-centred design pattern in HIMCoS is risky since it leads to overestimate the ability of the human operator who must perfectly behave when desired, within due response times, and who is also perfectly able to react facing unexpected situations: How can we be sure that he is able to realize all what he is intended to do and in the best possible way? And more, do his human reaction times comply with the high-speed ones of computerized artificial entities? Thus, what if he takes too much time to react? What if he makes wrong or risky decisions? What if he simply does not know what to do?

Moreover, one specificity in HIMCoS renders the techno-centred approach more risky. Indeed, as explained before, "bottom-up" emerging behaviours will occur in HIMCoS. Emerging behaviours are never faced (nor sought) in classical hierarchically/centralized control approaches in manufacturing. This novelty, analysed with regards to the need to maintain and guarantee especially the safety levels in manufacturing systems makes it more crucial. Typically, is the human operator ready to face the unexpected in front of complex self-organizing complex systems? This critical issue has seldom been addressed, see for example [12]. And, on the opposite point of view, what to do in case of unexpected events, for which no foreseen technical solution is available whereas the human is the only entity really able to invent one?

2.4 Why a so Obvious Assumption Remains Hidden?

From our point of view, three main reasons explain why this assumption remains hidden and is seldom explicitly pointed out.

The first one comes from the fact that researchers in industrial engineering are often not expert in or even aware of ergonomics, human factor or human-machine systems. A second one comes from the fact that integrating the human operator will require introducing undesired qualitative and blurring elements coupled to hardly reproducible and evaluable behaviours including complex experimental protocols potentially involving several humans as "guinea pigs" for test purpose. Last, the technological evolution in CPS, infotronics and information and communication technologies facilitates the automation of control functions (denoted LoA: level of automation), which make it easier for researchers to automate as much as possible the different control functions they consider.

For all these reasons, researchers, consciously or not, "kick into touch" or sidestep the integration of the human dimension, when designing their HIMCoS or their industrial control system.

3 Towards a More Human-Centred Control Design in IMS

In HIMCoS, and in industrial engineering in general, it is nowadays crucial to revise the basic research design patterns to adopt a more *human-centred* approach to limit the hidden assumption of the magic human. Obviously, this challenge is huge and complex but a growing number of researchers, especially in ergonomics and human-engineering address now this objective especially in the domain of industrial engineering. They typically work on the introduced LoA and the Situation Awareness as a prerequisite [13, 14]. Some EU projects have also been launched to illustrate this recent evolution (e.g., SO-PC-PRO "Subject Orientation For People Centred Production" and MAN-MADE "MANufacturing through ergonoMic and safe Anthropocentric aDaptive workplacEs for context aware factories in Europe").

This paper does not intend to provide absolute solutions but rather aims to set the alarm bell ringing and to provide some guidelines and insights for researchers to manage this hidden assumption of the "magic human" when designing the HIMCoS.

The main principle of a human-centred approach for the designer is to anticipate the functions that the human will operate, thus to determine the information he will need to understand the state of the system, to formulate a decision and last, to act. This anticipation must be accompanied by several main principles. Below is proposed a list that focuses on the decisional and informational aspects (for example, ergonomic aspects are not studied here, but can result in several items of the list).

The human can be the devil and unfortunately, some design engineers consider him as a devil: his rationality is bounded (cf. Simon's principle); he may forget, make mistakes, over-react, be absent or even be the root cause of a disaster. For example because of a bad understanding of the behaviour of a system that decides to switch to a secure mode (Three Mile Island), or acting bad because of a lack of trust in the automated system or a lack of knowledge about the industrial system (Chernobyl) [15]. The controlled system must be designed accordingly to manage this risk. A typical example of such a system is the one of the "dead-man's vigilance device" in train transportation where the conductor must trigger frequently a system so that the control system knows that he is really on command. This is a first level for a mutual control between the system and the human assuming that the other is of limited reliability. More elaborated levels would address the issue of wrong or abnormal command signals either from the human or the control system. This principle links then our discussion to safety (RAM: reliability, availability and maintainability) and FDI (fault detection isolation) studies, not only from technical point of view but also from human one.

The human can be the hero, and more often that we believe: he may save life using innovative unexpected behaviours (e.g., Apollo 13). The whole human-machine system must be designed to allow as much as possible the human to integrate unforeseen processes and mechanisms.

The human can be the powerless witness: He may be unable to act despite being sure he is right and the automated system wrong, for example, the spectator of a disaster due to the design of an automated but badly sized plant (Fukushima) [15]. The system must be designed to ensure its controllability by the human whenever desired.

The human is accountable, legally and socially speaking: the allocation of authority and responsibilities between human and machines is not so easy to solve (e.g., automatic cruise, air traffic control, automatized purchase processes for supply, automatic scheduling of processes, etc.). The designer must consider this aspect when designing and allocating decisional abilities among entities. In other words, if the human is accountable, he must be allowed to fully control the system.

Therefore:

The human must always be aware of the situation: According to Endsley [16], Situation Awareness (SA) is composed of three levels: SA1 (perception of the elements), SA2 (comprehension of the situation), SA3 (projection of future states). Thus each of these SA levels must be considered to ensure that humans can take decisions and make their mental models of the system evolve continuously (e.g., to take over the control or just to know what is the situation).

The LoA must be adaptive: some tasks must be automated and some others cannot be. But the related LoA must not be predefined and fixed forever. It must evolve according to situations and events, sometimes easing the work of the human (for example, in normal conditions) and other times, sending him back the control of critical tasks (for example, when abnormal situations occur). As a consequence, the control system must cooperate differently with the human according to situations: tasks allocation must be dynamic and handled in an adaptive way.

The diversity and repeatability of decisions must be considered, typically to avoid boring repetitive actions/decisions. This also requires to explicit as much as possible all the rare decisions for which the human was not prepared. For that, a time-based hierarchy (e.g., strategic, tactic and operational levels) and a typology of decisions (e.g., according to skill, rule or knowledge-based behavior) can be defined.

Therefore, **the human mental workload must be carefully addressed**: related to some of the previous principles, there exists an "optimal workload", between nothing to do, inducing potentially lack of interest and too much things to do, inducing stress and fatigue. A typical consequence is that the designer must carefully define different time horizons (from real time to long term), balance the reactions times of the human with the one of the controlled industrial system. This is one of the historical issues dealt with by researchers in human engineering [17].

4 Proposal of a Human-Centred Design Framework

For sure it is not possible to draw a generic model of a HIMCoS that complies for each possible case with all the previous principles. Despite this, we can propose a human-centred design framework to provide to researchers in IMS (and in more general, in industrial engineering) with some ideas to limit the magic human effect in their control system. For that purpose, Fig. 3 presents such a global framework. As suggested before, the process has been decomposed into 3 levels: *operational* for the short run, *tactical* at a higher hierarchical level for achieving the intermediate objectives and *strategic* at the highest level. The human may be apparently absent of the lower level, but this does not mean a fully automated system. We can therefore consider the automation in the system in 3 subsets as in nuclear plant

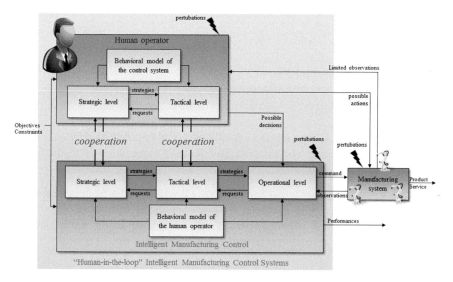

Fig. 3 Human-centred HIMCoS design approach

control: one subset is fully automated, a second one is not fully automated but the feedback experience enables to design procedures that the human must follow (a kind of automation of human), and the last subset is neither automated nor foreseen and therefore must be achieved thanks to the human inventive capabilities. This requires paying a particular attention when designing the whole system so that the humans are able to play the best of them especially when no technical solution is available!

This framework features some of the previously introduced principles. For example, a mutual observation (through cooperation) is performed to consider the limited reliability of either the human or the intelligent manufacturing control system. Also, different time horizon levels are proposed. But some other principles can be hardly represented in this figure. This is typically the case for the one dealing with the adaptive LoA. Research in this field is very active since few years. A famous guideline based on 10 levels has been proposed by [18], where at level 1, the control is completely manual while at level 10, the control is fully automatized. The 4th intermediary level corresponds to the DSS (the control selects a possible action and proposes it to the operator). At level 6, the control lets a limited time to the operator to counterbalance the decision before the automatic execution of the decision. This can be specified for each level (strategic, tactical, and operational). For example, it is nowadays conceivable that the Intelligent Manufacturing Control system depicted in Fig. 3 changes itself the operational decision level from a level 1–4 to the level 10 because of the need to react within milliseconds to avoid an accident while it lets the tactical decision level unchanged to an intermediary level. Researchers in automated vehicle addressed adaptive LoA, which may be inspiring in industrial engineering [19]. Works on Humans Machines Cooperation is one very promising track since the current technology allows embedding more and more decisional abilities into machines and transform them into efficient assistants (CPS, avatar, agents, holons…) to humans for enhancing performance. In such a context, it is suggested that each of these assistants embed:

- A Know-How (KH, knowledge and processing capabilities and capabilities of communication with other assistants and with the environment: sensors, actuators), and
- A Know-How to Cooperate (KHC) allowing the assistant to cooperate with others (e.g., gathering coordination abilities and capabilities to facilitate the achievement of the goals of the other assistants) [13].

Recent works have shown that team Situation Awareness can be increased when the humans cooperate with assistant machines equipped with such cooperative abilities. Examples were shown in several application fields: air traffic control, cockpit of the fighter aircraft and human robot cooperative rescue actions [20].

5 Conclusion

The aim of this paper was to raise awareness of the risk of maintaining hidden and true the "magic human" assumption when designing HIMCoS and at a more general level, industrial control systems with the human in the loop as a decision maker.

The suggested human-centred design aims to reconcile two apparently antagonist behaviours: the imperfect human, who can correct and learn from his errors, and the attentive and inventive human capable of detecting problems and bringing solutions even if they are difficult and new. With a human-centred design approach in IMS, human resources can be amplified by recent ICT tools to support them with decision and action. The integration of such tools leads to the question of the level of automation, since these tools could become real decision partners and even real collaborators for humans [21].

References

1. Cardin, O., Trentesaux, D., Thomas, A., Castagna, P., Berger, T., El-Haouzi, H.B.: Coupling predictive scheduling and reactive control in manufacturing hybrid control architectures: state of the art and future challenges. J. Intell. Manuf. doi:10.1007/s10845-015-1139-0 (2016)
2. Van Brussel, H., Wyns, J., Valckenaers, P., Bongaerts, L., Peeters, P.: Reference architecture for holonic manufacturing systems: PROSA. Comput. Ind. **37**, 255–274 (1998)
3. Leitão, P., Restivo, F.: ADACOR: a holonic architecture for agile and adaptive manufacturing control. Comput. Ind. **57**, 121–130 (2006)

4. Barbosa, J., Leitão, P., Adam, E., Trentesaux, D.: Dynamic self-organization in holonic multi-agent manufacturing systems: The ADACOR evolution. Comput. Ind. **66**, 99–111 (2015)
5. McFarlane, D., Giannikas, V., Wong, A.C.Y., Harrison, M.: Product intelligence in industrial control: theory and practice. Annual Rev. Control **37**, 69–88 (2013)
6. Lee, J., Bagheri, B., Kao, H.-A.: A cyber-physical systems architecture for industry 4.0-based manufacturing systems. Manuf. Lett. **3**, 18–23 (2015)
7. Gaham, M., Bouzouia, B., Achour, N.: Human-in-the-Loop Cyber-Physical Production Systems Control (HiLCP2sC): a multi-objective interactive framework proposal, service orientation in holonic and multi-agent manufacturing, pp. 315–325, Springer (2015)
8. Zambrano Rey, G., Carvalho, M., Trentesaux, D.: Cooperation models between humans and artificial self-organizing systems: Motivations, issues and perspectives. In: 6th International Symposium on Resilient Control Systems (ISRCS), pp. 156–161 (2013)
9. Oborski, P.: Man-machine interactions in advanced manufacturing systems. Int. J. Adv. Manuf. Technol. **23**, 227–232 (2003)
10. Mac Carthy, B.: Organizational, systems and human issues in production planning, scheduling and control. In: Handbook of production scheduling, pp. 59–90, Springer, US (2006)
11. Trentesaux, D., Dindeleux, R., Tahon, C.: A multicriteria decision support system for dynamic task allocation in a distributed production activity control structure. Int. J. Comput. Integr. Manuf. **11**, 3–17 (1998)
12. Valckenaers, P., Van Brussel, H., Bruyninckx, H., Saint Germain, B., Van Belle, J., Philips, J.: Predicting the unexpected. Comput. Ind. **62**, 623–637 (2011)
13. Millot, P.: Designing human-machine cooperation systems. ISTE-Wiley, London (2014)
14. Pacaux-Lemoine, M.-P., Debernard, S., Godin, A., Rajaonah, B., Anceaux, F., Vanderhaegen, F.: Levels of Automation and human-machine cooperation: application to human-robot interaction. In: IFAC World Congress, pp. 6484–6492 (2011)
15. Schmitt, K.: Automations influence on nuclear power plants: a look at three accidents and how automation played a role. Int. Ergon. Assoc. World Conf., Recife, Brazil (2012)
16. Endsley, M.R.: Toward a theory of situation awareness in dynamic systems. Hum. Factors: J. Hum. Factors Ergon. Soc. **37**, 32–64 (1995)
17. Trentesaux, D., Moray, N., Tahon, C.: Integration of the human operator into responsive discrete production management systems. Eur. J. Oper. Res. **109**, 342–361 (1998)
18. Sheridan, T.B.: Telerobotics, automation, and human supervisory control, MIT Press (1992)
19. Sentouh, C., Popieul, J.C.: Human–machine interaction in automated vehicles: The ABV project. In: Risk management in life-critical systems, pp. 335–350, ISTE-Wiley (2014)
20. Millot, P.: Cooperative organization for enhancing situation awareness. In: Risk management in life-critical systems, pp. 279–300, ISTE-Wiley, London (2014)
21. Millot, P., Boy, G.A.: Human-machine cooperation: a solution for life-critical systems? Work, **41** (2012)

Sustainability in Production Systems: A Review of Optimization Methods Studying Social Responsibility Issues in Workforce Scheduling

Carlos A. Moreno-Camacho and Jairo R. Montoya-Torres

Abstract Production scheduling in manufacturing systems is a highly complex task. It involves the allocation of limited resources (machines, tools, personnel, etc.) for the execution of specific jobs. In the case of workforce scheduling, unique human characteristics must also be considered, further complicating the task, as similar characteristics are not present in machines. Numerous published research works have examined issues of workforce scheduling by evaluating employee characteristics. Moreover, as business policies must nowadays sup-port social responsibility objectives, academic works have considered this dimension. This paper reviews academic literature on workforce scheduling strategies that consider social responsibility issues in order to identify quantitative methods and techniques employed. A systematic literature review is conducted to form an objective, rigorous, and reproducible framework that minimizes biases in the inclusion/exclusion of analysed works. Applications, trends, and gaps are identified, thus identifying pertinent avenues for future research.

Keywords Workforce scheduling · Manufacturing · Social responsibility · Modelling approaches · Systematic literature review

1 Introduction

Sustainability in manufacturing has become a hot topic in current research agenda due to customers' concerns about the impact that manufacturing activities have on both the environment and society. The goal is to simultaneously take into account economic, social and environmental performance metrics within the decision

C.A. Moreno-Camacho · J.R. Montoya-Torres (✉)
Escuela Internacional de Ciencias Económicas Y Administrativas,
Universidad de La Sabana, Bogota, D.C., Chía (Cundinamarca), Colombia
e-mail: jairo.montoya@unisabana.edu.co

© Springer International Publishing Switzerland 2016
T. Borangiu et al. (eds.), *Service Orientation in Holonic and Multi-Agent Manufacturing*, Studies in Computational Intelligence 640,
DOI 10.1007/978-3-319-30337-6_11

making process, known as the Triple Bottom Line (profit, planet and people) [6]. The academic literature has witnessed the appearance of several reviews on sustainable manufacturing, mainly focusing on the strategic decision-making levels: supply chain design, layout design, cleaner product and production mean design, construction, recycling process, etc. [4, 5]. As stated in [16], one of the main reasons for the strategic level emphasis is that much of the sustainability efforts have been driven by highest decision levels within organizations. According to [16], research considering sustainability issues, as a whole, at lower decision-making levels (i.e., operations control and scheduling) has been relatively limited. Some efforts have been made in some industrial settings by considering only the environmental dimension of sustainability [16]. To the best of our knowledge, the social dimension has been still less studied.

Moreover, at present, regulations in various countries and trade agreements across countries are increasingly addressing issues of social responsibility and employee wellbeing. The international quality standard ISO 26000:2010 "Guidance on Social Responsibility (SR)" recognizes labour practices as central to the formation of company SR policies. Such practices outline issues that organizations must address for their employees and subcontractors, by taking as a fundamental principles that personnel is not a commodity. Therefore, employees may not be treated as tools of production nor be subjected to the same market forces applicable to goods [9]. These guidelines assume the adoption of socially responsible labour practices that are fundamental to social justice, stability, and peace [9]. Hence, such aspects are pertinent for personnel scheduling in manufacturing and service organizations.

In this context, the aim of this paper is to review research works published from 2002 to 2014 that considers the social dimension of sustainability for manufacturing and service systems in which workforce and personnel resource scheduling is a central issue of concern. Indeed, workforce scheduling affects operating costs and customer service quality [1], while at the same time affects staff morale, mental health, social wellbeing and productivity [10, 11]. Given these considerations, labour is not merely a productive resource, as it is necessary to consider each employee as an individual with unique characteristics.

The goal is to identify at what extend social responsibility issues have been taken into account on workforce scheduling literature using optimization methods. This work will help advance knowledge on the application of international standards on social responsibility to personnel scheduling in manufacturing and service systems. A systematic literature review (SLR) approach is applied for the rigorous selection, inclusion and exclusion, and classification of articles to identify trends and gaps in scientific research and propose future lines of research. In turn, we illustrate how published academic works have incorporated employees as human beings in modelling approaches in workforce scheduling problems. The study period begins in 2002, as the ISO committee presented its report on the viability and convenience of delivering and international standard on social responsibility considering labour practices and employees' needs [8, p. 213].

2 Review Methodology and Framework to Classify Studies

The paper presents a systematic review of the academic literature on personnel scheduling. The five-step review methodology proposed in [2] is adopted for the identification, selection, and classification of studies and for the analysis and presentation of results. Unlike narrative reviews, this review approach, through the declaration of a searching process and establishment of exclusion/inclusion criteria, minimizes information bias, thus rendering this method systematic and reproducible [15]. Following are the explanation of how standard steps to carry out a systematic literature review were applied in the current paper:

Step 1 Question formulation: First, the main research question that will direct the literature review is defined. For the purpose of the current paper, we defined the following question: *At what extend quantitative modelling and resolution approaches have been applied to solve the workforce-scheduling problem considering social responsibility issues and human factors?*

Step 2 Study identification: This step involves selecting databases and search engines, as well as defining search criteria. In the current paper, the ISI Web of Knowledge/Science database was employed, as it offers academic institutions access to a broad collection of publications. Regarding the second criterion, documents published from 2002 to 2014 were collected. The following words and phrases were used to perform our search: Employee Timetabling, Multi-skilled Workforce, Labour + Scheduling, Employee + Scheduling, Workforce + Scheduling, Hierarchical Workforce Scheduling, Staff Scheduling, and Tour Scheduling Problem. Only journal papers were selected (i.e., editorials, letters to the editor, discussion articles, essays, and similar documents were excluded). This resulted in a total of 265 papers. We do recognize that there exist a lot of approaches on ergonomics studies and design methods, but we made to choice to exclude those topics from the current review.

Step 3 Study selection and evaluation: During this stage, the titles and abstracts of the 265 papers were reviewed to ensure relevance to the research question. A total of 93 papers were finally short-listed for further analysis. Of these, only 90 papers were examined in detail (two papers of the initial list were identified as literature reviews and one paper only presented a conceptual framework without providing solution methods).

Step 4 Analysis and synthesis: A total of 90 papers were read in detail and classified. The complete list of shortlisted papers is available at: http://jrmontoya. wordpress.com/research/workforce-scheduling-reviewed-papers-january-2015/. Papers were classified as follows:

- **Solution tools**: Employed quantitative tools.
- **Problem objectives**: Optimization objective(s).

- **Problem assumptions**: Various assumptions are presented about labour, demand, production shift management, vacation, etc. These were divided as follows:

  - *Constant labour*: The number of available employees is known at start of the planning horizon and remains constant throughout; hiring is not anticipated.
  - *Variable labour*: Employee hiring and firing during the planning horizon is allowed, due to fluctuations in demand between periods. Three forms of hiring are identified: full-time, part-time and hourly. This condition is associated with multiple shifts of variable lengths and start times.
  - *Homogeneous labour*: The paper defines this category as equality in productivity rates for all employees, regardless of the type of contract established.
  - *Heterogeneous labour*: Even when several employees offer the same skill, execution productivity rates for the same task are not equal.
  - *Hierarchical skills*: Employees have specific skills that allow them to complete certain tasks and that employees can be scheduled to perform lower-order tasks, but at the cost of their rank (for example, for an employee who possesses skills 1 and 2, if 1 is the more specialized skill, he/she can be scheduled to perform order-2 tasks at the cost of an order 1 employee).
  - *Non-hierarchical skills*: This condition recognizes that each employee has specific skills, but not all skills. Each employee can only be given tasks that he/she can execute, and therefore it is not possible to exchange one employee for another.
  - *Number of shifts*: Single or multiple shifts of variable or constant length can be allocated throughout the workday.
  - *Nature of demand*: Evaluates whether demand is deterministic or stochastic.

- **Production environment**: This item classifies papers based on the following production system configurations: flow-shop, job-shop, waiting lines and projects. We do recognize that some other production configurations exist and are studied in the scheduling theory; however, only those listed above are considered herein since they are the most complex ones found in short-listed papers.
- **Application**: Evaluates characteristics of the work presented and whether solution procedures are applied to case studies.

Step 5   Reporting of results: This paper presents for the first time the results of this review to the academic community. The following sections report the analysis of results and propose various research perspectives.

3 Findings

This section presents the main findings of the systematic literature review. Statistics about the number of papers published annually that meet our general search criteria and those that consider aspects of SR are shown in Table 1. An average of 7.1 articles were published per year, with the highest number of publications published from 2002 to 2006. Moreover, 24 % of the reviewed papers consider SR issues in the objective function such as workload balance, work stability, employee satisfaction and preference, ergonomic risk minimization, deviations in working day volumes, deviations in minimum required vacation days for a period of time, and maximum work hours. 4 % of the reviewed papers consider these criteria as soft constraints in the model, implying that a restriction breach does not cause infeasibility, but is penalized through the objective function. Examples include: assigning work to an unskilled employee, assigning working periods that exceed the specified maximum, anticipated employee scheduling before a period of rest is completed, among others. 49 % of the works discuss issues such as: employees with multiple skills, variations in employee productivity, employee assignment availability and fatigue. These considerations recognize employees as human beings rather than as mere productive resources.

In regard of the optimization objective, most commonly evaluated objective functions are minimizing production costs or labour costs (slightly more than 60 % of reviewed papers). Other objectives such as employee satisfaction (8 %), workload balance (7 %), and penalties for noncompliance with soft constraints (9 %) are also considered. As noted previously, some soft constraints that consider employee wellbeing can help solve related problems, and thus personnel satisfaction and productivity are equally as important as satisfying demand at reasonable costs [14].

Regarding the problem solution technique, classical Operations Research methods such as mathematical programming, heuristics and meta-heuristics are employed to solve the problems. Binary variables are employed when the requirement is to assign employees to certain tasks based on shifts of variable start times and length while respecting the working hour maximum, among other conditions. Hence, mixed-integer linear programming (MILP) modelling is the most widely employed solution method (47 % of the short-listed papers). Heuristic methods were the second most frequently identified solution technique (24 % of reviewed papers), which are used in instances where MILP capacities are limited. Among the reviewed papers, 20 % of them use decomposition methods calculating

Table 1 Number of papers that include SR criteria in model formulations

Location of SR criteria	Time period				
	2002–2006	2007–2010	2011–2014	Total	Percentage
Objective function	5	10	8	23	25
Soft constraints	0	1	3	4	4
Model conditions	12	14	20	46	49

the number of employees needed to meet shift labour needs and then determining workdays and days off for each employee. Enumeration algorithms (Branch & Bound, Branch & Price, Branch & Cut) and methods of Column Generation have also been proposed. Meta-heuristic procedures such as Genetic Algorithm (GA), Tabu Search (TS), Simulated Annealing (SA) and Particle Swarm Optimization (PSO) are most frequently employed.

Another interesting issue to evaluate in this review were the job conditions as described in Sect. 2 (Step 4). The most common scenarios for scheduling are: constant labour involving homogeneous and non-hierarchical skills conducted over multiple shifts per day (an employee can only work one shift per day, but can work different shifts during the scheduling horizon), and constant demand or known demand at the start of the planning period. These conditions can be applied effectively to multiple industrial manufacturing sectors. Other conditions such as variable labour, multiple workplaces, heterogeneous skills and variable demand appear primarily related to the service industry (medical services, postal services, check-in counters, maintenance services, and call centres, among others). As a matter of fact, 55 % of shortlisted papers studied an actual application to real-life manufacturing or service industry, while the other 45 % address theoretical problems with solution procedures tested using random-generated data sets.

The configuration of production system was another pertinent criterion for paper classification. Manufacturing personnel scheduling problems are often modelled to ensure that a sufficient number of employees meet daily work requirements several studies do not specify the system configuration. As a consequence, it was possible to classify only 26 % of reviewed papers. The most frequently used configurations correspond to queuing models (12 % of papers). Flow shop and job shop configurations are respectively studied in 7 % and 3 % of reviewed papers. Applications include manufacturing facilities and postal service activities. Moreover, resource-constrained project scheduling (RCPS)-based models (4 % of papers) are common in sectors such as construction, as one company is responsible for several projects.

4 Overview of Further Research Opportunities

Despite the development of extensive studies and production scheduling automation software, personnel and staff scheduling remains of great interest to both academics and practitioners. As noted above, this has occurred because unique company and employee characteristics presuppose the construction of a model with different conditions and all the more with the gradual inclusion of assumptions that consider employees as human beings. It is worth highlighting how, in recent years, social responsibility criteria have been considered by the academic research, including: heterogeneous labour, fatigue, ergonomic hazards, job training, learning curves, etc. Although these features further complicate the problem, they are of relevance to industry.

In addition, most of the reviewed papers aim to solve tangible issues, suggesting that organizations are invested in resolving production-scheduling issues in ways that reconcile employee work- and family-life responsibilities. This benefits the employee, but it also benefits the company through higher productivity and service quality [7, 10–12]. This is even more relevant in sectors where high staffing turnover rates affect profits of businesses that incur constant costs due to hiring, training, and employee development [3, 17].

From these findings, as noted in [13], a major research question remains unsolved: what considerations about labour practice social responsibility help to solve personnel scheduling problems to minimize errors resulting from overlooking employee variability?

Based on these findings and our literature review analysis, we propose that future research considers one or more of the factors listed hereafter:

1. Heterogeneous labour with productivity rates that are either stochastic or deterministic but variable as a function of time, with implicit features such as learning curves, work monotony, and employee fatigue during work shifts.
2. Aspects of family and social order that affect work employee performance and cause re-scheduling due to absenteeism.
3. Model evaluation with multiple objectives through which employee satisfaction and conciliation of work and family life are considered. Some of these considerations have been studied in [13, 17].
4. The development of social welfare programs and evaluations of their effects on the loyalty, morale, health and productivity of employees; strategic programs that produce better outcomes for the development of operational level solutions.
5. The development of efficient tools for solving tangible personnel scheduling problems that considers personnel. These may involve heuristics, meta-heuristics, or computer simulation models for stochastic process modelling.

5 Concluding Remarks

This paper presented an updated literature review that examines various dimensions of personnel scheduling with a particular emphasis on issues linked with the practice of social responsibility. Also, this paper intended to evaluate how these factors are considered in current scheduling research.

Our review showed that within the context problem features, rather than objective functions, labour practices that consider unique features that distinguish employees from other productive resources are accounted for most often. Mixed Integer Programming (MIP) is the most widely used approach for problem modelling and resolution. Decomposition techniques are also very often employed; however because of its complexity, these techniques do not guarantee global optimality for the integrated problem. Likewise, we highlight the development of heuristic algorithms and the effective application of meta-heuristics such as Particle

Swarm Optimization (PSO) and evolutionary strategies. Results reported in the literature show that these procedures outperform other meta-heuristics such as Tabu Search (which is commonly used to solve related scheduling problems).

Regarding objective functions, the present review shows that while cost and labour minimization approaches remain the most widely used, additional objectives seek to maximize employee satisfaction based on their work day, shift, vacation, workload and stability preferences. In regard of applications, call centres, hospitals, airport terminals, and postal organizations are most commonly studies as actual case studies, as well as the manufacturing industry. This is attributable to the fact that industrial sectors have developed various strategies to face demand changes (e.g., safety inventory levels, response times), while service sectors seek to mitigate such changes through strategies of capacity, resulting in coordinated personnel scheduling.

Overall, the present review shows how academic studies related to quantitative models elucidate issues of personnel scheduling that involve social considerations and human factors. Regarding the importance of these criteria in solving this problem, standard parameters for the inclusion of such issues in the problem formulation are not identified or established due to industry differences and varying scheduler assessments. However, the classification identifies aspects of social responsibility in labour practices addressed in the literature.

Acknowledgment The work presented in this paper was supported under a postgraduate scholarship awarded to the first author by Universidad de La Sabana.

References

1. Alfares, H.K.: Survey Categorization, and comparison of recent tour scheduling literature. Ann. Oper. Res. **127**(1–4), 145–175 (2004)
2. Denyer, D., Tranfield, D.: Producing a systematic review. In: Buchanan, D.A., Bryman, A. (eds.) The Sage Handbook of Organizational Research Methods, pp. 671–689, Sage Publications Ltd (2009)
3. Florez, L., Castro-Lacouture, D., Medaglia, A.L.: Sustainable workforce scheduling in construction program management. J. Oper. Res. Soc. **64**(8), 1169–1181 (2013)
4. Garetti, M., Taisch, M.: Sustainable manufacturing: trends and research challenges. Prod. Plann. Control **23**, 83–104 (2012)
5. Gunasekaran, A., Spalanzani, A.: Sustainability of manufacturing and services: Investigations for research and applications. Int. J. Prod. Econ. **140**, 35–47 (2012)
6. Montoya-Torres, J.R.: Designing sustainable supply chains based on the triple bottom line approach. Proc. 2015 Int. Conf. Advanced Logistics and Transport (ICALT 2015). Valenciennes, France, 1–6. May 20–22 (2015)
7. Musliu, N., Gärtner, J., Slany, W.: Efficient generation of rotating workforce schedules. Discrete Appl. Math. **118**(1–2), 85–98 (2002)
8. Navarro García, F.: Responsabilidad Social Corporativa: Teoría y práctica (2da ed). ESIC Editorial (2008)
9. Organización Internacional de Normalización.: ISO 26000 Guía de Responsabilidad Social, Of 2010. Geneve, Switzerland (2010)

10. Petrovic, S., Van den Berghe, G.: A comparison of two approaches to nurse rostering problems. Ann. Oper. Res. **194**(1), 365–384 (2012)
11. Puente, J., Gómez, A., Fernández, I., Priore, P.: Medical doctor rostering problem in a hospital emergency department by means of genetic algorithms. Comput. Ind. Eng. **56**(4), 1232–1242 (2009)
12. Rocha, M., Oliveira, J.F., Carravilla, M.A.: A constructive heuristic for staff scheduling in the glass industry. Ann. Oper. Res. **217**(1), 463–478 (2014)
13. Thompson, G.M., Goodale, J.C.: Variable employee productivity in workforce scheduling. Eur. J. Oper. Res. **170**(2), 376–390 (2006)
14. Topaloglu, S., Ozkarahan, I.: Implicit goal programming model for the tour scheduling problem considering the employee work preferences. Ann. Oper. Res. **128**(1–4), 135–158 (2004)
15. Tranfield, D., Denyer, D., Smart, P.: Towards a methodology for developing evidence-informed management knowledge by means of systematic review. Br. J. Manage. **14**(3), 207–222 (2003)
16. Trentesaux, D., Prabhu, V.: Sustainability in manufacturing operations scheduling: stakes, approaches and trends. In: Grabot, B., Vallespir, B., Gomes, S., Bouras, A., Kiritsis, D. (eds.) APMS 2014, IFIP AICT 439, pp. 106–113. Springer, Heidelberg (2014)
17. Wright, P.D., Mahar, S.: Centralized nurse scheduling to simultaneously improve schedule cost and nurse satisfaction. Omega **41**(6), 1042–1052 (2013)

Identifying the Requirements for Resilient Production Control Systems

Rengarajan Srinivasan, Duncan McFarlane and Alan Thorne

Abstract Tighter supply chains create an increasing need for manufacturing organisations to become more flexible and more able to cope with disruptions drives the need for resilient production. Further, the interconnected nature of production environment and the complexities associated with the adoption of lean and process automation requires monitoring and control system to have increasing functionalities. Beyond simple monitoring and control, production systems are required to analyse information from disparate sources, detect abnormal deviations and then to react and cope with those deviations in a more effective manner. In this paper key requirements for resilient production systems are developed by establishing the links between production disruption and the required resilient capabilities. This then translates into requirements for resilient control and tracking in production systems.

Keywords Resilient production · Control · Tracking · Disturbance handling

1 Introduction

The complex dynamics associated with manufacturing enterprises requires resilient properties. Those manufacturing organisations are faced with disruptions ranging from natural events (such as floods, hurricanes, earthquakes), transportation disruptions (such as road closures) and internal disruptions (such as quality issues,

R. Srinivasan (✉) · D. McFarlane · A. Thorne
Department of Engineering, Cambridge University, CB3 0FS Cambridge, UK
e-mail: rs538@cam.ac.uk

D. McFarlane
e-mail: dm114@cam.ac.uk

A. Thorne
e-mail: ajt@cam.ac.uk

© Springer International Publishing Switzerland 2016
T. Borangiu et al. (eds.), *Service Orientation in Holonic and Multi-Agent Manufacturing*, Studies in Computational Intelligence 640,
DOI 10.1007/978-3-319-30337-6_12

125

resource breakdowns, material delivery issues) [1]. The ability to identify, respond and cope with disruptions is becoming essential for these firms to operate in a global environment and be competitive at the same time.

The intricate nature of manufacturing systems coupled with interdependent processes and human interactions are placing additional constraints on management and control. The dependence of manufacturing systems on automation and their ability to provide a higher degree of state awareness is important to identify the onset of disruptions and also to develop resilience strategies [2].

From a manufacturing point of view, the need for resilience arises from the fact to know the operational status of the systems in real time, to be aware of the state so as to identify the onset of disruptions. Additionally, there is also the need to determine or infer from the state, the most appropriate course of action that will either reduce the impact of disruption or allowing coping with it. Essentially, this implies manufacturing needs efficient tracking of information and control systems that can infer from the tracking information to determine/act on the required mitigation strategy.

Resilience in general can be defined as the ability of the system to cope with unexpected changes [2]. For production systems, resilience is closely associated *with robustness, responsiveness* and *agility*. Robustness is the ability of the production system to maintain its goal or the desired output in the face of disturbances [3]. Responsiveness is defined as the ability of the production system to respond to disturbances [3]. On the other hand, agility refers to quick and adequate changes to disturbances.

Despite the clear need for resilient systems in manufacturing, there has been a lack of understanding of the key requirements that are needed for a resilient production control. The key objective is to link disruption analysis to the design of a resilient production control strategy. Additionally, the proposed approach is demonstrated in a lab used as an experimental facility.

2 Identifying the Requirements of Resilient Production System

The need for resilient production system stems from the fact that production operations are inherently prone to various disturbances and therefore the key enabler for resiliency is the ability to avoid/survive/recover from disturbances. Consequently, disturbance identification and their characteristics will influence the resilience capability requirements, which will then lead to establishing the resilience strategies and the consequent control and tracking requirements that will enable the production system to be resilient to the identified disturbances. This process is illustrated in Fig. 1.

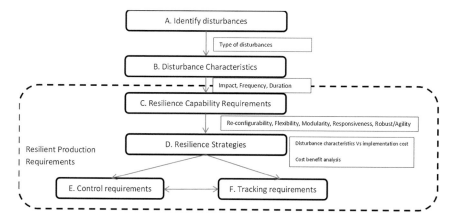

Fig. 1 Process for identifying resilience requirements

2.1 Disturbance Identification

In order to understand the needs of a resilient system, it is essential to develop deeper understanding of the nature of the disturbances and their associated impact on production goals. This forms the preliminary step for identifying the requirements of resilient production systems. Actual and potential disturbances affecting the production system need to be considered. The typical disturbances that occur in manufacturing are [3] upstream disturbances (e.g. material quality, supplier delivery), internal disturbances (resource breakdowns, quality, operator errors, material handling errors) and downstream disturbances (demand fluctuations).

2.2 Disturbance Characteristics

The consequences of disturbances and their impact on production system need to be evaluated. The characteristics of a particular disturbance such as frequency of occurrence and duration will have varying consequences on the effectiveness of achieving the production goals. Additionally, the impact of disturbance may be localised or distributed, and can also propagate through the system [3]. These disturbance characteristics will have an influence on the requirements of resilient production control.

The analysis of the various disturbances in the production system will provide a clear assessment of the current production capacities and flexibilities, while indicating the key areas that lack responsiveness and thus resilience. This will allow prioritising the main disturbances to be handled by the production system to become resilient.

2.3 Resilient Capability Requirements

For a production system to be resilient it needs to have the capability to:

- Detect/Diagnose/Assess: Related to disturbance identification.
- Recover/Adapt/Cope: Related to the ability to react/act to mitigate the impact of disturbance.
- Anticipate/Learn: Related to the ability of predicting the disturbance and learn from past experiences.

Resilient production system should have the capability to detect the disturbance in a timely manner: In order for the production system to be resilient, it is essential to detect the occurrence of disturbance in a timely manner, which will then enable quick response to mitigate or recover. The detectability and recognition capability require real-time dynamic information related to the production system. This capability is related to the responsiveness of the system. Additionally, manufacturing operations involving human operators needs right information regarding the occurrence and type of disturbance at the right time such that the response and recovery time are minimised.

Resilient production system should have the capability to be flexible and/or have redundancy: Resilient properties of production systems are generally enabled through the system flexibility and by having operational redundancy and inventory redundancy [4]. The redundancy can be incorporated into the production system by having additional resources, capacity and/or flexible resources [5]. In order to react or cope with disturbances, the production system should be flexible. The type flexibility needed will be governed by the nature of disturbance and will in general include resource, material handling, routing and process flexibilities.

Resilient production system should be adaptable/reconfigurable: In order to react to changing market dynamics and external demand fluctuations, rapidly reconfigurable production systems are needed. This requires key capabilities such as modularity of hardware and software and integrability to add new resources and future technologies.

2.4 Resilience Strategies

Disturbance analysis and the resilience capability requirements are used to develop strategies for utilising the underlying response capability in the system to cater for disturbances. The resilience strategies should align with the following:

- Resilience strategies should be aligned with the phases of disruption. The resilience phases indicate the timing of implementing the strategy. This implies the strategy can be implemented before (avoidance), during (survival) or after (recovery) the occurrence of disturbance.

- Resilience strategies should determine when and how to utilise the dynamic capability of the system. This is determines the way in which the inherent flexibility should be utilised for mitigating the occurrence of disturbances. For example, this includes determining the buffer sizes and location, and inspection station locations.

2.5 Track and Trace Requirements

The key requirements for handling disturbances from a tracking perspective are the need to update the operational status (awareness) and the ability to detect the occurrence of disturbance.

Tracking system should be able to capture and process information from various production resources. The need for tracking from a resilience perspective gives rise to certain characteristics. The tracking system should be able to capture and sense the data at an aggregate level, relating to the behaviour of a processing line rather than the individual machines or products [6]. Similarly, the data should be processed and analysed at higher level by combining information from people, product and resource [6]. Additionally, the tracking system should be able to capture event related information rather than raw data. This should allow the tracking system to have additional functionalities, moving from data logging to recording event related data.

Tracking system should be able to communicate the required information for various production resources. The tracking system should be able to manage the information by communicating the message required to the right entity and also to store the data in a meaningful manner. Particularly for resilient production systems, there is a requirement for automatic capture of real time data on uniquely identified products, providing visibility of operations by associating products with their current location, condition/status and history [7]. Additionally, it is also important to capture process parameters, resource data and associate them with products.

Tracking system should be interoperable. In order to capture data from disparate sources, it is essential to consider existing standards and issues related to interoperability. Tracking in production set up needs to combine data from physical resources and control systems; it is thus important to integrate this information in a seamless manner. Standards on communication and data representation must be considered.

2.6 Resilient Control Requirements

The role of production control is to interface with planning and schedule, and to execute the respective operations based on the schedule. On the other hand, for resilient production control, it should also determine and/or anticipate deviations and to make necessary control adjustments accordingly.

Control should be able to communicate real-time information and incorporate them in analysis and decision making. There is an inherent need to capture the information from the control system and communicate that information to the tracking system and wider business entities. This allows production system to gain operational visibility for disturbance handling. Additionally, the control system should be able to incorporate information from disparate sources (through tracking system) to analyse and act on the information signals.

Control system should be able to infer current state (local and global) and predict/identify the onset of disturbances. For resilient production systems, it is essential for the control system to know the operational state for the purpose of detecting the disturbances. Also, the control system should be able to forewarn or predict the occurrence of disturbances.

Control should have the ability to react/control for handling or coping with disturbances. In addition to detecting disturbances, the resilient control system should dynamically react or cope with disturbances. In this aspect the following requirements are identified:

Control should be de-centralised: Centralised control will become complex and difficult to adapt for handling disturbances [8]. Therefore, distributed intelligent control will be more suitable for resilient production systems. In order to be resilient, the distributed control should be product or resource-based.

- In resource-based architecture, the set of resources is able to allocate jobs without a centralised support, allowing the system to be flexible and reconfigurable [8].
- In product-based architecture, the customer's order/product drives the production process by negotiating with the individual resources. This allows the system to cope with variations in customers' preferences and customisation.

Control should be adaptable (self-organising) and utilise dynamic capabilities as needed: In order to cope with disturbances, the control system should be adaptable and utilise the flexibilities provided by planning, processes, resources and operational flow. Additionally, the system should be re-configurable and therefore the control system should be self-organising.

3   Illustration Using Laboratory Production System

As an example, we consider the production of a gearbox. The gearbox consists of
(a) a metal casing consisting of two parts, top and base; (b) a plastic top cover
(c) gears, which go into the casing. The metal casing is machined by a 5-axis *CNC*
machine and the plastic covers are made by a vacuum forming machine. The lay of
the production system consists of three cells. Cell 1 is the main manufacturing cell,
where the metal cases and plastic covers are formed. Cell 2 is a sub-assembly cell
which assembles the plastic top with the associated metal top. Cell 3 is the final
assembly cell, where the gears are assembled and fastened, as represented in Fig. 2.

3.1   Disturbance Analysis

The key disturbances that occur in this production systems are

- *Part Quality Problem*: Due to the variations in the forming process, the plastic
  parts have deformity. These variations cause additional re-work and also causes
  significant delay in fulfilling customer orders. Additionally, since there were no
  buffer stock, any disruption occurring has high impact on the performance of the
  production system.
- *Material mishandling*: Disturbances also occur during material transfer, where
  human operators tend to misplace the parts on the trolleys. When the robots starts
  the assembling sequence, due to parts in wrong location, different sub-assemblies
  are combined together and this again caused delay in final delivery. Furthermore,

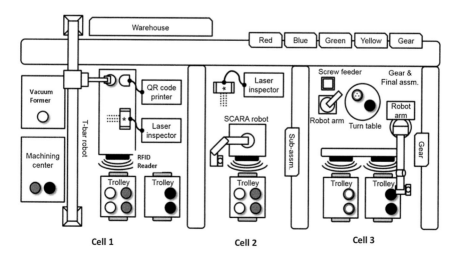

Fig. 2   Production layout

the manual operators moved wrong trolley to workstations, thereby causing delays in the production schedule.

3.2 Resilience Capability Requirements

- *Capability to detect*: In order to detect the part quality variations, inspections stations were added in cell 1 and cell 2. In order to mitigate the wrong delivery parts and trolley, direct part marking with unique ID's was added.
- *Capability to be flexible*: For handling part quality losses, additional buffers are added in cell 1 and cell 2. The buffer sizes can vary depending on the order profile and quality losses. To identify parts, the robots have embedded data matrix reader.

3.3 Resilience Strategies

Inspection information is used to decide whether the part quality is achieved. If the parts are failed, then buffers are utilised to re-assign parts to orders. Unique ID on parts helps in associating/disassociating parts to orders. To handle part misplacements, the robot's data matrix reader are used to read the unique ID before the start of operations, thereby eliminating the possibility of wrong assembly of products.

3.4 Track and Trace Requirements

- *Capture and process information*: Inspection information is captured along with process parameters and is associated with uniquely identified parts. Unique identification of parts is provided by data matrix and trolley identification is provided by UHF *RFID*. Trolley location information is captured at each cell to identify the location and the trolley type. Data matrix readers capture the unique ID of each part and are associated with the trolley ID and location.
- *Communicate*: The tracking system communicates the trolley arrival events at each cell to the control system. Process parameters and inspection information are transferred to the tracking system by the control system. The latter will update the information systems, providing real time visibility of operations and orders.
- *Interoperable*: The tracking system using standardised interfaces to communicate with the devices. The Data matrix and *RFID* readers use standard TCP/IP messages and the control system communicates via PLC through the network.

3.5 *Resilience Control Requirements*

- *Information Handling*: Control system analyses the part-id of each part before proceeding with the operation. Additionally, the control system is transferring real-time information regarding the status of orders/parts and resources.
- Operational State Awareness: The control system is communicating regularly with the tracking system before and during operations, thus enabling it to determine the current state and detect the occurrence of disturbances.
- *React/control*: Distributed *holonic* control principle is implemented, where during quality disturbance the orders dynamically allocate parts through negotiation with other order parts and buffers. Resources also co-operate with orders/parts to carry out the job sequences.

4 Conclusions

In this paper, we derived the requirements of resilient production systems, by combining disturbance analysis and resilience capability requirements. The design of resilient systems needs to broaden existing approaches by incorporating new thinking, as well as incorporating better human/system interaction and complex interdependencies of distributed control systems [9]. Integrating new information gathering and processing techniques with intelligent distributed control systems will enable resilient production systems that can detect and cope with disturbances in more dynamic and efficient manner. Further investigations will be conducted to evaluate the requirements based on extending the case example implementation.

References

1. Hu, Y., Li, J., Holloway, L.: Resilient control for serial manufacturing networks with advance notice of disruptions. IEEE Trans. Syst. Man Cybern. **43**, 98–114 (2013)
2. Rieger, C., Gertman, D., McQueen, M.: Resilient control systems: next generation design research. In: 2nd Conference on Human System Interactions, HSI'09, pp. 632–636, May 2009
3. Matson, J., McFarlane, D.: Assessing the responsiveness of existing production operations. Int. J. Oper. Prod. Manage. **19**(8), 765–784 (1999)
4. Hu, Y., Li, J., Holloway, L.: Towards modeling of resilience dynamics in manufacturing enterprises: literature review and problem formulation. In: IEEE International Conference on Automation Science and Engineering, pp. 279–284. CASE 2008, Aug 2008
5. Zhang, W., van Luttervelt, C.: Toward a resilient manufacturing system. CIRP Ann. Manuf. Technol. **60**(1), 469–472 (2011)
6. McFarlane, D., Parlikad, A., Neely, A., Thorne, A. (2012): A framework for distributed intelligent automation systems developments. In: Borangiu, T., Dolgui, A., Dumitrache, I., Filip, F.G. (eds.) 14th IFAC Symposium on Information Control Problems in Manufacturing, vol. 14, pp. 758–763 (2012)

7. Brintrup, A., Ranasinghe, D., McFarlane, D.: RFID opportunity analysis for leaner manufacturing. Int. J. Prod. Res. **48**(9), 2745–2764 (2010)
8. Bussmann, S., McFarlane, D.: Rationales for holonic manufacturing control. In: Proceedings of the 2nd International Workshop on Intelligent Manufacturing System (1999)
9. Rieger, C.: Notional examples and benchmark aspects of a resilient control system. In: 3rd International Symposium on Resilient Control Systems (ISRCS), pp. 64–71, Aug 2010

Requirements Verification Method for System Engineering Based on a RDF Logic View

Albéric Cornière, Virginie Fortineau, Thomas Paviot
and Samir Lamouri

Abstract Requirements Engineering (RE) is often seen as a preliminary phase to design, however in a PLM (Product Life-cycle Management) context its range widens to the whole life cycle of a product. Verification of requirements is one of the activities associated with RE, that consist in asserting the actual system complies to the requirements. This verification is usually performed manually, one requirement at a time, relying on the engineer's expertise. In order to enable automatic verification of requirements on large and complex systems, we propose a semantic model of requirements based on business concepts and modelled with RDF (Resource Description Framework). This model joins the logical and the concrete views on the system in a twin network of RDF triples. The application domain is the nuclear industry.

1 Introduction

Requirement Engineering (henceforth RE) is often considered a preliminary process to the product life cycle, aimed at formalizing a consistent set of specifications from the stakeholders goals. The very term "Requirements Engineering" was introduced by a technical note in the US Defense and Space Systems Group. From the nineties onwards it became a research field with its own subset of activities related to the RE process: elicitation, modelling, validation and verification of the requirements set, and management of the requirements and their traceability [14]. RE is thus

A. Cornière (✉) · V. Fortineau · T. Paviot · S. Lamouri
LAMIH Arts et Métiers ParisTech, 151 bd de l'hôpital, Paris, France
e-mail: alberic.corniere@ensam.eu

V. Fortineau
e-mail: virginie.fortineau@ensam.eu

T. Paviot
e-mail: thomas.paviot@ensam.eu

S. Lamouri
e-mail: samir.lamouri@ensam.eu

© Springer International Publishing Switzerland 2016
T. Borangiu et al. (eds.), *Service Orientation in Holonic and Multi-Agent Manufacturing*, Studies in Computational Intelligence 640,
DOI 10.1007/978-3-319-30337-6_13

135

intimately bound to the semantic representation of the system; the requirements verification consists in the confrontation of the "logical" view in system engineering (henceforth SE) with the current design in a "product" view.

In a Product Life Cycle Management (PLM) approach, RE has a wider range, though: it becomes an iterative process involved at each phase in the product life, at each definition level of the system [2] to specify, dimension, build or maintain a requirements-compliant system. A first proposal of a generic model for automatic verification of requirements has already proposed in [6]. This generic model's elements themselves represent business objects from the plant [5]. This work's issue is modelling the logical view of the system, of its elements, and of their semantic interactions with requirements, in order to effectively lever the proposed conceptualisation and to verify requirements automatically. The goal is to propose a requirements modelling workshop in the information system that allows to SE engineers to verify compliance to requirements automatically from within a PLM suite. In this contribution, we propose a model for the "logical" view of the system. This model relies on defined business concepts and on a representation of requirements through RDF (Resource Description Framework) triples (see Sect. 3). As "information is in the relations" [12], the triple (a relation between two given concepts) is the fundamental element of the proposed paradigm. The second keystone to this contribution is a two-parts formalization: on one hand the semantic business objects (as "patterns") and on the other hand the concrete data, designated as "occurrences". A case study is presented (Sect. 4) to illustrate the match of requirements defining elements and objects of the system, and to illustrate the associated verification process. Finally the advantages of the proposed model are discussed in Sect. 5, as are the semantic and technical perspectives.

2   State of the Art

2.1   Modelling and Verification of Requirements in a SE Work Flow

Most contributions on RE are either from the software engineering field, or focus on the activity of requirements elicitation [1, 11]. A complete literature review on the different methods for elicitation has also been proposed in [8]. Even regarding requirements on products or services, research on requirements verification has led to specific algorithms for manual verification methods, one requirement at a time, generally on the basis of textual statements [3, 13]. Another part of literature is generally designated as artefact-Based RE and focuses on classifying the requirements prior to their verification. Proposed models are thus focused on requirements management and traceability. However no conceptual formalization is proposed in these works, which generally settle for textual statements, as is the case for ISO15288 or in the SysML formalism. For instance, Berkovich et al. [4] notes that a limit to the RD-Mod method is the absence of semantic link between the requirements list (a document) and the functional architecture of the product. Yet, a necessary condition for automatically verifying the compliance of a product to requirements is a reliable semantic formalization of the requirements, linked to the functional and organic system definition, and associated to a generic model. This work is then not about the elicitation of requirements, which are considered as input. It does not contribute to the process of requirements management within a work flow either. This contribution presents a generic model for requirements elicited beforehand, that allows to automatically verify compliance to a full set of requirements through reasoning on the logical view of the product, while ensuring traceability throughout the life cycle of the product. In order to do so, the requirement model's genericness, the system representation's semantic richness, and the mappings between them are crucial.

2.2 Representation of the System in System Engineering

Expressing of the requirements relies on an abstract view of the system, representing different granularity levels. Known in SE as a "logical" view of the system, it complements the "product" view, which consists in a faithful representation of the actual system. For instance, for a system engineer, a requirement can apply to all valves in a system while another applies to valves in subsystem of this supply system. In the logical view then, valves of the system and of the subsystems will be two different concepts, though actually the physical valve $valve_{328}$ is both a valve in the system and its subsystem. To one object in the product view, correspond several in the logical view. The requirements model must then be relying on a logical model of the plant. This logical model holds in a same whole a multilevel, multi-view, abstract representation of a system. In this aspect it essentially consists in a labelled, directed multi-graph, interconnecting system taxonomies, functions, components, etc. It must also exhibit a sufficient semantic richness to express any and all requirements, which is not the case for existing models, most notably for existing standards [7, 9].

2.3 Constraints Due to Complexity and to the PLM Approach

Traceability: For the traceability to be reliable in an automated process, the information associated with a requirement must be readily computable for the verification process in addition to being usable by the system engineers. In order to identify and trace what led to this requirement, and in order to automatically verify its satisfaction, a rich semantic is needed, that allows a distinction among the different relations and concepts involved in the definition of requirements.

Automation of the verification process: given the scale and the complexity of the considered systems, relying on human expertise alone to verify the requirements is not sufficient. An automatic verification process can resolve a large number of requirements, saving the engineers expertise for cases where it is most needed.

Atomic requirement verification: the complexity and the scale of a nuclear power plant induces a potentially arbitrary large, interconnected logical network. The information needed to verify each requirement must be reduced as much as possible in order to limit analysed data, and thus to obtain results in an acceptable time frame.

Reasoning reliability: in the context of nuclear engineering, reliable processes and results are crucial. Some requirements are related to safety or nuclear security, for which it is necessary for the verification to be as reliable as possible.

Genericness of the model: a generic model is chosen for requirements, in order to both use a project-agnostic syntax for requirements definition, and to use a same algorithm for the verification of any requirement, avoiding to resort to specific methods and algorithms for different cases.

2.4 Conceptual and Generic Representation of Requirements for Verification

A conceptual and generic model of requirements for verification has been proposed and discussed in [5] and is presented on Fig. 1. It holds the concepts involved in a requirement, but it is not implementable as it is. This conceptualization relies on 5 generic elements: first the circumstantial **conditions**, in which the requirement constrains an **attribute** of a **constrained element**, which can be a function, a system or a component. To comply with the requirement, the attribute must be consistent with a set of admissible values: the **criterion**. Verification of a requirement's satisfaction then consists in the comparison of the actual values of the

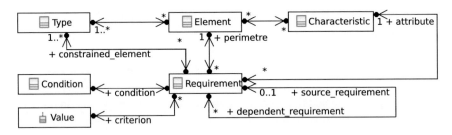

Fig. 1 Requirements related classes for automated verification in UML syntax

attribute to the criterion. To avoid any ambiguity in the lookup for all the constrained elements of a given requirement, contextual information are added as the
perimeter of a requirement. For instance "The valves of the cooling system must be
manoeuvrable locally" requires the "*valves*" as a **constrained element** to present
the **attribute** "*manoeuvre type*" equal to the **criterion** "*local*". It does not concern
every valve in the plant though, only those in its **perimeter**, "*the cooling system*".

3 Modelling Requirements from a Networks of Occurrences Defined by Semantic "Patterns"

A network data defined as a network of patterns: The defining elements of a
requirement are by essence to be applied on the occurrences (individuals) of the
product view, according to concepts of the logical view. The dataset on which they
rely is then a rich multi-view, multilevel representation: it consists in a network in
which concepts, taxonomies, states of the system, etc. are all represented by network nodes; their relations to one another represent the different business views of
the system. From this diversity of individuals and of their relations comes the
semantic richness needed to match requirements to the elements in the product
view. Chains and sub-graphs of relations in this network act as "patterns" for the
definition of requirements: "the valves of the cooling system" for instance are
matched by the pattern "*individual—of type—valve*, where *individual—is in—
cooling system*".

On Fig. 2, requirement 4 states that function_1 must be achieved in a duration
under 1 h. Two parts of the RDF logical view are clearly visible, with a "data" part
and a part for information and "patterns". The RDF graph is represented in black.

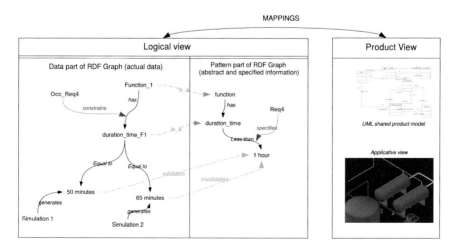

Fig. 2 Twin graphs representing the conceptual and factual data of the system

On the left, function_1 has it own duration, while on the right a function "may have" a duration as a characteristic. The expression of requirement 4 is in red and has two parts: the constraint "data" i.e. the *constrained element—has—attribute* triple, and the pattern specification for *attribute—less than—1 h*. Finally the verification in green is the comparison of the actual value from the data side with the specification. In this example two values from different simulations are considered, one satisfies the requirement while the other does not.

RDF Syntax: As seen above, the logical view for the system is represented as a network of relations between individuals. RDF is well suited to declare individuals and their relations as triples, in the form "*item_1—relation—item_2*", which allows to represent a directed, labelled multi-graph.

Generic modelling of the requirements from a twin network of triples: The requirements can be described as RDF triples as well, a first triple in the form "*constrained element—has—attribute*" and a second one in the form "*attribute—is within—criterion*" being the defining pattern for verification itself. Using RDF as a mean to map this expected state of the design allows to distinguish within the data model between the specification and the actual state of affairs, while the capability to process them together remains. In terms of abstraction level, the requirement is effectively a bridge from the specification (a conceptual representation) to the reality of the design or implementation (a specific occurrence of the system).

Method for automatic verification of requirements: Verification occurrences can then be derived from the requirements definitions as triples, by matching the triples against sub-views of the logical view (i.e. views that consider only the relations "has" from a concept to an attribute, and "is within" from an attribute to a criterion). This generic "inspector" consists in a simple comparison and traceability meta-data: one is created and run for each of the occurrences in the product view that match the *constrained element* defining pattern. Their individual and aggregated results in turn determine global satisfaction of the (conceptual) requirement as a whole.

4   Case Study

The case study is a work in progress drawn from the field of nuclear plants engineering, a field characterised by systems of a great complexity and of a large scale. Semantic elements necessary to the representation of a nuclear plant are several billions, not counting the abstract objects for instance representing system groups of phases of life. Also, the complexity of a nuclear installation arises from the multiple non-trivial interactions of its elements, tangibly in the involved physical processes, as well as in the data model representing them. By this two respects, nuclear engineering is representative of complex, large-scale system engineering. Consider a requirement from this application case (illustrated in Fig. 2), "Safety functions must be performed in less than an hour". To model this requirement, we use the defining elements presented in 1. It constraints the **attribute** *"duration time"* of its **constrained element** *"safety function"* in its **perimeter**: the *"plant"* to be inferior to *"one hour"*, its **criterion**. Triples representing this requirement in the conceptual domain reflect these elements: A first triple of concepts denotes the constrained element and its attribute: *"safety function—has—duration time"*, and another triple references the specification with *"duration time—less than—one hour"*. On the data side of the graph, as soon as an individual is referenced as a safety function (through the *'is_a'* relationship from this data occurrence to the *safety function* concept), an occurrence for the requirement is created and the triple *"Occ_ReqN—constraints—Function_1"* is created. *Occ_ReqN* is the entry point for the verification algorithm, which reports failure if *Function_1* does not have a *duration time* attribute, or if it's out of admissible range. In turn, the *ReqN* individual in the concepts database reports failure along with the faulty occurrence, if any of the occurrences derived from it report failure. Every verification is made separately on data of each constrained element for traceability. In essence, verification itself consists in the creation of a *requirement verification occurrence* for each constrained element, which is an entry point for a comparison of this element attribute to the requirement's.

5 Discussion and Perspectives

The model and methods proposed in this contribution have limits to take into account. First and foremost they rely on a consistent semantic representation of the actual data to verify. It can be problematic, if possible at all, to retrieve these data into the logical view if they are created or altered in specific environments, such as specialized CAD software. This interoperability problem has been treated and discussed in [10]. Overcoming this issue can be done with semantic mappings between the global logical view and the various product views. Such mapping demands reasoning capabilities, as presented in the perspectives below. The vast and thorough logical view also raises an issue related to scale: as the system's data grows in number of relation, the complexity of the mapping with patterns grows exponentially, potentially leading to arbitrary long processing. As of this contribution no assumptions were made on the structure of the logical network; which may present properties that could be used to optimize processing on the network for allocation. The *perimeter* is another possible mitigation of the scale. As for the verification itself, as the verification model being both light and generic [6], it can be processed in batches efficiently. The model and the methods described in this model being based on a formalism designed for data exchange and storage, using it not only across the life cycle of the product, but also from a project to a following one in the same family of products, is a foreseeable perspective. Storing requirements templates in the knowledge base could allow to generate conceptual requirements from the knowledge and rules contained in it—possibly through rules application. Reasoning directly on the "patterns" and product network directly is not trivial, not only due to the size of the graph, but mainly because of its complexity and the variety of rules that need to be applied on it to make it consistent. e.g. transitivity of the *system—has—subsystem* relationship has to be performed. In practice, completion of the information requires many additions to the network. Ontologies can be leveraged, taking benefit from their reasoning capabilities and support of SWRL rules to explicit the implicit part of information, prior to generating a more complete RDF graph of individuals.

6   Conclusion

In this study, we propose a semantic representation of knowledge associated with a system's design and of its definition data. The first contribution of this model is the representation of data and knowledge in a generic by design, labelled directed graph constructed with RDF triples. The second contribution of this model is the twin formalization of this graph allowing to separate specification from system data. The proposed models can be used as an information source for automatic requirements verification, as well as a definition for requirements template and type patterns for system elements.

References

1. Ahmed, N., Matulevicius, R.: A method for eliciting security requirements from the business process models. In: CAiSE Forum and Doctoral Consortium, pp. 57–64 (2014). url http://ceur-ws.org/Vol-1164/PaperVision08.pdf
2. Arnold, S.: ISO 15288 Systems Engineering System Life Cycle Processes. International Standards Organisation (2002)
3. Ben-David, S., Sterin, B., Atlee, J.M., Beidu, S.: Symbolic model checking of product-line requirements using sat-based methods. In: IEEE/ACM 37th IEEE International Conference on Software Engineering (ICSE), 2015, vol. 1, pp. 189–199. IEEE (2015)
4. Berkovich, M., Leimeister, J.M., Hoffmann, A., Krcmar, H.: A requirements data model for product service systems. Req. Eng. **19**(2), 161–186 (2014)
5. Cornière, A., Fortineau, V., Paviot, T., Lamouri, S., Goblet, J.L., Platon, A., Dutertre, C.: Modelling requirements in service to PLM for long lived products in the nuclear field. In: Advances in Production Management Systems. Innovative and Knowledge-Based Production Management in a Global-Local World, pp. 650–657. Springer, Berlin (2014)
6. Cornière, A., Fortineau, V., Paviot, T., Lamouri, S.: Towards a framework for integration of requirements engineering in PLM. In: 15th IFAC Symposium on Information Control Problems in Manufacturing INCOM 2015. IFAC-PapersOnLine **48**(3), 283–287 (2015). doi http://dx.doi.org/10.1016/j.ifacol.2015.06.095.   url   http://www.sciencedirect.com/science/article/pii/S2405896315003341
7. ISO: Iso 10303 (2014)
8. Nisar, S., Nawaz, M., Sirshar, M.: Review analysis on requirement elicitation and its issues. Int. J. Comput. Commun. Syst. Eng. (IJCCSE) **2**, 484–489 (2015)
9. OMG: Sysml v 1.3 (2012). http://www.omg.org/spec/SysML/1.3

10. Paviot, T.: Méthodologie de résolution des problèmes d'interopérabilité dans le domaine du product lifecycle management. Ph.D. thesis, Ecole Centrale Paris (2010)
11. Rahman, M., Ripon, S., et al.: Elicitation and modeling non-functional requirements-a pos case study (2014). arXiv preprint arXiv:1403.1936. url http://arxiv.org/pdf/1403.1936
12. Tsuchiya, S.: Improving knowledge creation ability trough organizational learning. In: Proceedings of International Symposium on the Management of Industrial and Corporate Knowledge, ISMICK, Compiègne, France (1993)
13. Viriyasitavat, W., Da Xu, L.: Compliance checking for requirement-oriented service workflow interoperations. IEEE Trans. Ind. Inform. **10**(2), 1469–1477 (2014)
14. Zave, P., Jackson, M.: Four dark corners of requirements engineering. ACM Trans. Softw. Eng. Methodol. (TOSEM) **6**(1), 1–30 (1997)

Approaching Industrial Sustainability Investments in Resource Efficiency Through Agent-Based Simulation

F. Tonelli, G. Fadiran, M. Raberto and S. Cincotti

Abstract To develop more sustainable industrial systems industrialists and policy makers need to better understand how to respond to economic, environmental, and social challenges and transform industrial behaviour by leveraging appropriate industrial technology investments to reshape the current manufacturing value chain. Investments have to be collected on the private as well as public sides taking into account the stakeholders' macro-economic framework. Since aggregate mathematical models, assuming informed, rational behaviour leading to equilibrium conditions cannot catch the resulting complexity, an agent-based modelling and simulation approach is proposed to investigate policies to support investments in resource efficiency. The EURACE agent-based framework has been adopted and modified by coupling the environmental sector with other established macroeconomic dimensions. The findings of this research establish the potential and capability of the proposed approach for investigating policies for sustainability transition analysis and evaluation.

Keywords Industrial sustainability · Resource efficiency investments · Economics · Agent based model

1 Industrial Sustainability and Resource Efficiency Investments

The modern global industrial system has delivered major benefits in wealth creation, technological advancement, and enhanced well-being in many aspects of human life. However, industry is estimated to be responsible for some 30 % of the greenhouse gases (GHG) in industrialized countries and is a major consumer of primary resources [1], and resource scarcity and resultant price and supply issues

F. Tonelli (✉) · G. Fadiran · M. Raberto · S. Cincotti
Department of Mechanical Engineering, Energetics, Management and Transportation
(DIME), Polytechnic School, University of Genoa, Genoa, Italy
e-mail: flavio.tonelli@unige.it

© Springer International Publishing Switzerland 2016
T. Borangiu et al. (eds.), *Service Orientation in Holonic and Multi-Agent Manufacturing*, Studies in Computational Intelligence 640,
DOI 10.1007/978-3-319-30337-6_14

require new strategies and innovation at different levels [2]. Although some progress towards sustainability has been achieved (i.e. eco-efficiency, cleaner production, recycling initiatives, and extended producer responsibility directives), overall sustainability at the macro-level has not improved. Similarly, despite impressive improvements in material productivity and energy efficiency in many industry sectors, overall energy and material throughput continues to rise. Even in the instance where an industrial firm-level innovation seems to be highly effective, the system-level effectiveness of most currently proposed models on the current manufacturing value chain is largely unproven, and the long-run implications for sustainability are poorly understood even because of lack in measuring these effects [3] or supporting decision making [4]. To develop more sustainable industrial systems industrialists and policy makers need to better understand how to respond to economic, environmental, and social challenges and transform industrial behaviour accordingly by leveraging appropriate industrial technology investments to reshaping the current manufacturing value chain. Investments have to be collected on the private as well as public sides taking into account the stakeholders' macro-economic framework. In order to identify effective incentives and enablers to leverage, at least partially, financial capital investments into sustainability, the dynamic interactions between financial capital, natural resources and technology have to be analysed in their interdependences, increasing the overall complexity. Several articles have focused attention on attaining successful levels of sustainability with resource efficiency coupled with minimal impact on governments, firms, households, research and public players, such as Meyer et al. [5], Behrens et al. [6], Millok et al. [7], and Söderholm [8] as summarized in Table 1.

Table 1 Main stakeholders' options towards industrial sustainability transition

Government	Firm	Households	Research and public
Material tax Energy tax Sales tax Profit tax Subsidy in low pollution technologies	Capital and consumption goods production Natural resources-material mining Resource efficiency (material and energy) Product development (i.e. DfX, eco-design) Production technology investments (i.e. cleaner production, pollution prevention) Pricing Cost-savings valuation and accounting Competition and sales	Consumption goods choice option Price change decisions Long lasting, modular and updatable products preference End-of-life conscious management	Sustainable scenarios and case studies analysis Environmental regulations Sustainability and waste measurements Attainable reduction targets setting Economic versus industrial modelling Policy tests and guidance

2　Why Agent-Based Model and Simulation Approach?

Agent-based modelling (ABM) has gained prominence through new insights on the limitations of traditional assumptions and approaches, as well as computational advances that permit better modelling and analysis of complex systems and particularly in the sustainability domain [9]. Agent-based models in the industrial sustainability field are emerging and various authors have identified the potential value and effectiveness and advocated such simulation approaches. Bousquet et al. [10] provide a review of multi-agent simulations and ecosystem management, Trentesaux and Giret [11] the adoption of manufacturing holons towards sustainable manufacturing operations control. Monostori et al. demonstrated the possibility of ABM integrating manufacturing sustainability through multi-agent systems [12] while Davis et al. used ABM integrated with a life cycle assessment to investigate effects of energy based infrastructure system on its environment [13]. Yang et al. used an agent-based simulation approach to investigate economic sustainability to evaluate waste-to-material recovery system [14]. Typically, such works have focused on specific environmental issues such as carbon, or waste, and generally this is modelled at the individual firm level. Cao et al. demonstrated agent interactions between the factory, consumers and the environment focusing on eco industrial parks [15]. The findings of this research established the potential and capability of ABM for investigating decision options for optimal eco-industrial systems in both 'open-loop' and 'closed-loop' structures/systems.

In summary, ABM is an interesting emerging field that seems to have significant potential for exploring the complex issues of transitions towards industrial sustainability at system level. Applied to the next generation of industrial systems and manufacturing value chains, such models have the potential to give industrialists a needed test-bed for safe, low-cost management and policy experiments.

The reported research aims at using agent-based modelling to simulate the dynamic models of the capital investment in technological investments for promoting industrial sustainability. The dynamic model adopted is the EURACE agent-based framework [16–18], a platform demonstrating the potential for modelling of large complex systems with many thousands of heterogeneous agents. The EURACE platform has been modified by coupling the environmental sector with other established macroeconomic dimensions, by integrating material input, resource efficiency and environmental policy considering explicitly the extracting and manufacturing phases.

3　Developed Model and Assumptions

Eurace is an agent-based macroeconomic model and simulator, which is under development since 2006 thanks to the funding of two different European projects. In the original version, Eurace agent population is characterized by different types of agents: households, which act as workers, consumers and financial investors; consumption goods producers (CGPs), henceforth firms, producing a homogenous consumption goods; a capital goods producer; commercial banks and two policy makers agents, namely a government and a central bank, which are in charge of fiscal and monetary policy, respectively. Agents interact through different markets where consumption and capital goods, labour and credit are exchanged in a decentralized setting with disperse prices set by suppliers and based on costs. Agents' behaviour is modelled as myopic and characterized by limited information, scarce computational capabilities and adaptive expectations. For instance, CGPs are characterized by a short-term profit objective and make production and investment plans where expected future revenues are based on backward-looking expectations determined by past sales and prices. In particular, production plans depend on past sales and the inventory stock, along the lines of the inventory management literature, while sale prices are determined by a mark-up on costs (wages and debt interests). Investment plans depend on the cost of capital goods and the present value of the additional foreseen revenues, but are limited by both by internal and external financing capabilities.[1] Households set the consumption budget out of their income following a wealth to income target ratio, according to the theory of buffer-stock saving [20]. Savings can be allocated in stocks (i.e. the claims on firms/banks equity and future dividends) and government bonds, which are traded in a centralized Walrasian financial market. Banks have the function to provide short-term loans to firms at an interest rate determined by the cost of central bank loans, i.e. the policy rate, plus a mark-up. It is worth noting that, in line with the working of the banking system in a modern capitalist economy [21], banks lending is not limited by the available liquidity and, whenever a bank grants a loan, a corresponding deposit, entitled to the borrower, is created on the liability side of the bank's balance sheet. Bank lending is however limited by a Basel II-like capital requirement rule; in this respect, each bank assesses the loan risk by considering the financial leverage of the prospective borrower before deciding about a loan request.

In order to address the issue of industrial sustainability, the Eurace model has been enriched with the following features:

1. Beside labour and capital, raw materials are a new required input of production for CGPs, whose production technology is now characterized by two nested production functions, namely, a Leontief with two inputs, raw materials on one side and, on the on the other side, labour and capital, which are coupled according to the usual Cobb-Douglas technology. Therefore, differently form labour and capital, raw materials are a non-substitutable production factor. The amount of raw materials necessary to produce 1 unit of consumption goods is called *resource intensity*, while its inverse is the *resource efficiency*. Resource intensity/efficiency is heterogeneous across CGPs and decreases (increases) over time according to the vintage of capital goods of any CGP.

---

[1]The pecking order theory [19] is adopted to determine a hierarchy of financial sources for the firm.

2. A new agent, called "mining company", which extracts raw materials to be sold to CGPs. We assume for simplicity that the raw materials price is exogenously given and that there are no extractions costs. Therefore, revenues and profits of the mining company coincide. Profits are paid out as dividends to the mining company's shareholders, which can be partially or totally the households populating the Eurace economy. Raw materials costs can therefore be "recycled" partially or totally back into the Eurace economy.

3. Capital goods are characterized by a resource efficiency parameter whose value depends on the time capital goods are produced and delivered to CGPs. In particular, the parameter's value increases according to an exogenously given yearly growth rate IR. The vintages of capital goods owned by each CGP set their resource efficiency.

4. The government levies a new tax, called *environmental* or *material tax*, which is applied to each CGP and computed as a percentage of the value of the raw materials input. A the same time, the government subsidizes CGP's investments by rebating a percentage of their capital goods expenses, up to the amount of environmental taxes paid (restricted case—S1) or without limitations (unrestricted case—S3),

5. Increasing resource efficiency and its related saving in raw materials costs and environmental taxes as well as subsidies are taken into account by CGPs in their net present value calculation to decide investments in new capital goods.

4 Simulations Validation, Results, and Discussion

This section provides views from simulation experiments through a set of snapshots of revenue recycling effects and efficiency investment decisions with consideration of environmental tax and subsidy. The research deals with subsidy receipts on capital costs for environmental tax paying firms only at 9 environmental tax levels on raw material/production input (only one level will be reported on this paper version while for the complete experimental set can be accessed at the following dropbox link[2]). The experimental simulation is carried out over a sample period of 20 years (240 months).

Figures 1, 2 and 3 compare two levels of recycled percentage of mining company's earnings to revenue into Eurace economy, to which 100 % (ER1) produced better results over 50 % (ER05). For a simpler representation, the figures are only produced for one selected environmental tax rate of 2.5 %, in reference to one subsidy percentage (5 % restricted-S1 while the unrestricted-S3 has not been reported) and efficiency dynamics (IR0 and IR2).

Results show that the higher the recycled earnings, the better the system is. Increase in material consumption, GDP and employment levels are observed in

---

[2]https://www.dropbox.com/sh/e052vgaix6a1x87/AADjTOSe0oRe4vcuoHcyKxCQa?dl=0.

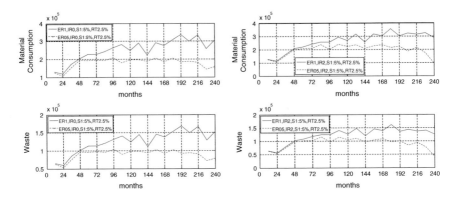

Fig. 1 Material consumption and waste levels (S1: IR0, IR2, ER1 vs. ER05)

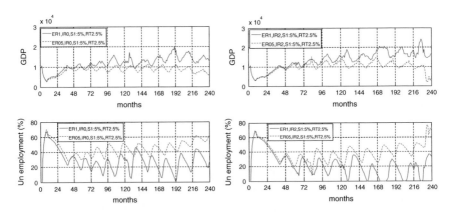

Fig. 2 GDP and unemployment (S1: IR0, IR2, ER1 vs. ER05)

both efficiency dynamics. That is, with or without efficiency investment gains, the system improves with increasing recycled earnings.

For further validation of efficiency investment dynamics and distinguished difference, the next section focuses more on 20 years averaged performance of 9 environmental tax levels between efficiency investment decisions (partial results presented because of space limitations[3]).

For observations, Figs. 3, 4, 5, 6 and 7 display trends between the old (IR0) and new efficiency investment gains (IR2) system. A target level of 2 % annual efficiency gains was used in relation to an observed average annual resource productivity growth rate for EU27 members. This is focused on the case of 100 % recycled earnings, in reference to the best validated performance (see Figs. 1 and 2).

---

[3]**NB**: IR0 (straight lines) versus IR2 (dash lines)
Subsidy percentages: 0 %-red; 5 %-blue; 10 %-pink; 15 %-black; 20 %-green.

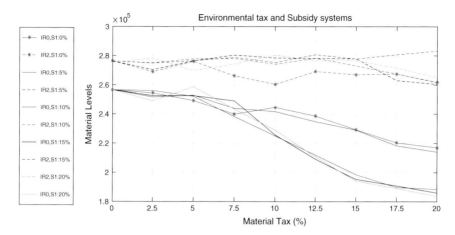

Fig. 3 Material consumption levels (IR0 vs. IR2)

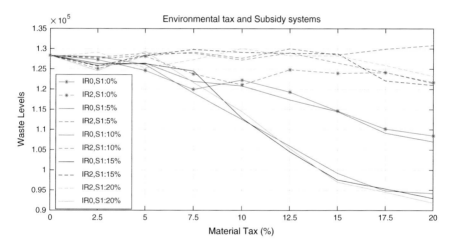

Fig. 4 Waste levels (IR0 vs. IR2)

From the figures, it is evident that systems with efficiency investments produced higher material consumption. Figure 4 shows the inverse proportion of waste emission levels which are higher under efficiency gain systems due to suspected rebound effects of higher material consumptions. On the positive side, Fig. 4 shows the reduced margin in waste levels between inefficiency system (IR0) and efficiency gain system (IR2), indicating an improvement in waste gap release. Furthermore the critical decline in material (Fig. 3) and waste levels (Fig. 4) at higher material tax rates indicates the powerful effect of environmental policy tool for reducing consumption of a targeted good, but on the verge of drop in employment (Fig. 6) and GDP (Fig. 7).

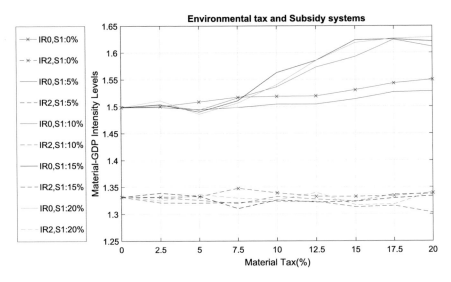

Fig. 5 Intensity levels (IR0 vs. IR2)

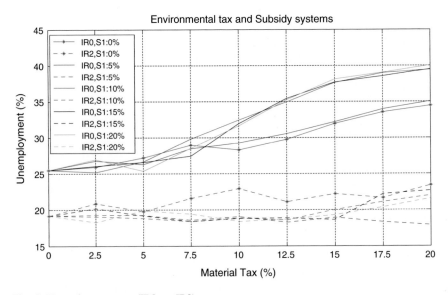

Fig. 6 Unemployment rate (IR0 vs. IR2)

For intensity, the lower the value the better the system, although lower values may sometimes indicate a presence of poor economic performance. Concerning intensity, efficiency system gain produces mostly lower trends at all levels, suggesting a better overall system (Fig. 5). Generally the introduction of subsidy payments proves to be effective with minimizing shock effects.

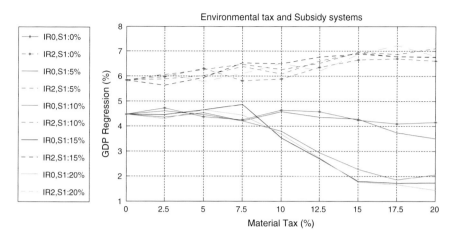

Fig. 7 GDP growth rate (regressive) (IR0 vs. IR2)

Finally, a presence of tax shift is observed as government adjusts general tax in response to different environmental tax levels and subsidy. A similar finding was observed in [22] since material and GDP levels may increase following events of material tax levies or may be due to environmental tax impact on firm production and competition levels, as debated by Conrad, that higher environmental taxes may reduce output significantly and weaken the job creation market, and thereby enforce firms to engage in abatement investments and input subsidies [23].

5 Conclusions

The paper contributes to agent based modelling towards sustainability transition; simulations confirm that environmental tax acts as another source of revenue, reduction of material consumption and environmental policy. Results from a comparison of IR0 versus IR2 are provided in Table 2. Summarizing, environmental taxes alone are not the best way for attaining selective sustainability, but rather, in combination with incentive packages for firms to invest in efficient production techniques, as both the Government and the Industry key players in achieving sustainability goals. Evidently, firm and government actions may harm sustainability transition when adopted without proper monitoring and regulation programs. The paper also shows notable shortcoming related to study approach by using an open-loop framework. As this initial study focuses only on efficiency

Table 2 Results comparison between IR0 and IR2

NO resource efficiency investment gain (IR0)	YES resource efficiency investment gain (IR2)
System is more sensitive to increasing environmental taxes; more evident at excessive levels, that is, above 10 % tax rates	In general, the system is less sensitive to increasing environmental taxes. More evident under restricted subsidy method. Increasing
Increasing environmental taxes significantly reduces material consumption but on the verge of inflicting employment and GDP; more notable after 10 % tax levy and irrespective to subsidy method or percentage	environmental taxes produces different reactions according to subsidy payment method. Increasing taxes not damaging to the economy, especially with restricted subsidy, but significantly on a downward spiral with
Tax shifts are more evident at higher environmental tax levels of restricted subsidies	higher unsustainable rates of unrestricted subsidy system in a similar pattern to
Higher intensity level, but with no particular trend to environmental tax increases	inefficiency investment (IR0)
Increasing environmental taxes is significantly damaging to the economy. Inclusion of subsidy did not display significant improvement	Inclusion of low subsidy aids system improvement. Higher levels should be cautionary as associated with increasing general taxes
Subsidy payments arguably not significantly effective enough	Tax shifts are more evident between unrestricted subsidy levels
	A probable presence of rebound due to increased material consumption, including while increasing environmental taxes
	Lower intensity level, but with no particular trend to environmental tax increases

comparison, the next stage will consider a better focus on subsidy method comparison and impact trend. Other considerable future features include introduction of recyclable product and multiple material input options.

Acknowledgments The authors acknowledge EU-FP7 collaborative project SYMPHONY under grant No. 611875. A special thanks to Dr. William Samuel Short, postdoctoral from Institute for Manufacturing, Cambridge University for precious contribution and collaboration on industrial sustainability suggestions with respect to agent-based modelling and simulation.

References

1. Evans, S., et al.: Towards a Sustainable Industrial System (2009). Available at: http://www.ifm.eng.cam.ac.uk/sis/
2. Tonelli, F., Evans, S., Taticchi, P.: Industrial sustainability: challenges, perspectives, actions. Int. J. Bus. Innov. Res. **7**(2), 143–163 (2013)
3. Taticchi, P., Tonelli, F., Pasqualino, R.: Performance measurement of sustainable supply chains: a literature review and a research agenda. Int. J. Prod. Performance Manage. **62**(8), 782–804 (2013)
4. Taticchi, P., Garengo, P., Nudurupati, S.S., Tonelli, F., Pasqualino, R.: A review of decision-support tools and performance measurement and sustainable supply chain management. Int. J. Prod. Res. 1–22 (2014)

5. Meyer, B., Distelkamp, M., Wolter, M.I.: Material efficiency and economic-environmental sustainability. Results of simulations for Germany with the model PANTA RHEI. Ecol. Econ. **63**(1), 192–200 (2007)
6. Behrens, A., Giljum, S., Kovanda, J., Niza, S.: The material basis of the global economy: worldwide patterns of natural resource extraction and their implications for sustainable resource use policies. Ecol. Econ. **64**(2), 444–453 (2007)
7. Millock, K., Nauges, C., Sterner, T.: Environmental taxes: a comparison of French and Swedish experience from taxes on industrial air pollution. CESifo DICE Rep. J. Inst. Comparison **2**(1), 30–34 (2004)
8. Söderholm, P.: Taxing virgin natural resources: lessons from aggregates taxation in Europe. Resour. Conserv. Recycl. **55**(11), 911 (2011)
9. Thomas, A., Trentesaux, D.: Are intelligent manufacturing systems sustainable? In: Borangiu, T., Trentesaux, D., Thomas, A. (eds.) Services Orientation in Holonic and Multi-agent Manufacturing and Robotics, pp. 3–14. Springer International Publishing, Berlin (2014)
10. Bousquet, F., Le Page, C.: Multi-agent simulations and ecosystem management: a review. Ecol. Model. **176**(3–4), 313–332 (2004)
11. Trentesaux, D., Giret, A.: Go-green manufacturing holons: a step towards sustainable manufacturing operations control. Manuf Letter, To appear (2015)
12. Monostori, L., Váncza, J., Kumara, S.R.T.: Agent-based systems for manufacturing. CIRP Ann. Manuf. Technol. **55**(2), 697–720 (2006)
13. Davis, C., Nikolić, I., Dijkema, G.P.J.: Integration of life cycle assessment into agent-based modeling. J. Ind. Ecol. **13**(2), 306–325 (2009)
14. Yang, Q.Z., Sheng, Y.Z., Shen, Z.Q.: Agent-based simulation of economic sustainability in waste-to-material recovery. In: 2011 IEEE International Conference on Industrial Engineering and Engineering Management, pp. 1150–1154 (2011)
15. Cao, K., Feng, X., Wan, H.: Applying agent-based modeling to the evolution of eco-industrial systems. Ecol. Econ. **68**(11), 2868–2876 (2009)
16. Cincotti, S., Raberto, M., Teglio, A.: Credit money and macroeconomic instability in the agent-based model and simulator Eurace. Economics: The Open-Access. Open-Assess. E-J. **4** (2010-26) (2010)
17. Cincotti, S., Raberto, M., Teglio, A.: Part II Chapter 4: The EURACE macroeconomic model and simulator. In: Aoki, M., et al. (eds.) Complexity and Institutions: Markets, Norms and Corporations, Masahiko Aoki, Kenneth Binmore, Simon Deakin, Herbert Gintis, pp. 81–104. Palgrave Macmillan (2012)
18. Raberto, M., Teglio, A., Cincotti, S.: Debt deleveraging and business cycles. An agent-based perspective. The Open-Access. Open-Assess. E-J. **6**, 2012-27 (2012)
19. Myers, S., Majluf, N.: Corporate financing and investment decisions when firms have information investors do not have. J. Financ. Econ. **13**(2), 187–221 (1984)
20. Deaton, A.: Household saving in LDCs: credit markets, insurance and welfare. Scand. J. Econ. **94**(2), 253–273 (1992)
21. McLeay, M., Radia, A., Thomas, R.: Money creation in the modern economy. Bank Engl. Q. Bull. **54**(1), 14–27 (2014)
22. Ekins, P., Pollitt, H., Summerton, P., Chewpreecha, U.: Increasing carbon and material productivity through environmental tax reform. Energy Policy **42**, 365–376 (2012)
23. Conrad, K.: Taxes and subsidies for pollution-intensive industries as trade policy. J. Env. Econ. Manage. **25**(2), 121–135 (1993)

Part IV
Holonic and Multi-Agent System Design for Industry and Services

Increasing Dependability by Agent-Based Model-Checking During Run-Time

Sebastian Rehberger, Thomas Aicher and Birgit Vogel-Heuser

Abstract Agent-oriented software engineering (AOSE) is a paradigm for distributing intelligent control mechanisms (ICM) within an automated production system (aPS). Benefits resulting from AOSE have been surveyed in many applications as route-finding, plug-and-produce techniques and also in the control of Smart Grids. To ensure safe functionalities, i.e. dependability or uptime, of distributed technical systems for instance by conducting simulation, virtual commissioning, the execution of test cases and model-checking are commonly investigated in aPS during the design phase. In this paper we analyze an automatic diagnostic method to increase dependability by using model-checking during run-time, based on discretized models of the mechanical plant as well as models of the PLC software. Consequently the algorithm is incorporated into a software agent and logically coupled to a particular aPS module. Thus, the dependability for introducing novel product types, which have not been involved in the design process, could be increased. The evaluation of our approach is shown at a small lab-scale production system by searching for counter-examples of combinations with control actions and work piece (WP) types with modified mass, that may lead to a production halt.

Keywords Automation production systems (aPS) · Agents · Model-checking · Modelling · Verification

S. Rehberger (✉) · T. Aicher · B. Vogel-Heuser
Institute of Automation and Information Systems, Technische Universität München, München, Germany
e-mail: rehberger@ais.mw.tum.de

T. Aicher
e-mail: aicher@ais.mw.tum.de

B. Vogel-Heuser
e-mail: vogel-heuser@ais.mw.tum.de

© Springer International Publishing Switzerland 2016
T. Borangiu et al. (eds.), *Service Orientation in Holonic and Multi-Agent Manufacturing*, Studies in Computational Intelligence 640,
DOI 10.1007/978-3-319-30337-6_15

1 Introduction

In recent years the introduction of techniques described as cyber-physical systems (CPS) became quite common. One of the main contributions of such systems is to increase autonomous behaviour of automated production systems (aPS). These systems are divided into modules containing mechanics, electrical/electronic and involve local intelligence by a distributed control. Dependability is still a major factor for ensuring uptime and preventing unplanned production interruptions. To increase flexibility of an aPS, new design paradigms, e.g. agent-based software-engineering (AOSE) or service-oriented architectures (SOA) were developed. The main goal is to avoid disadvantages of monolithic control architectures by locating the intelligent control algorithms logically or physically at distributed aPS modules. To accelerate the implementation of novel technologies in industry, verification methods such as model-checking, to ensure safe functionality, e.g. dependability, of the control software, its hardware and the controlled plant, are required. The goal of formal verification by model-checking is to ensure that the system's discrete states and its concurrences with the controller's states will never lead to a fault or potential damaging combination.

For application of model-checking during the design process, the boundaries for the discrete behaviour of the plant are defined. In order to use automation modules in a long term, model checking must be re-applied in a manner involving the adaption of the plant and further proving whether the defined specifications still hold. In this paper we approach this circumstance by applying model-checking online at run-time, based on a proactive propagation of the new product properties by an agent. The paper is structured as follows: First we will give a state-of-the-art of the application of model-checking in the automation domain and further basics of AOSE in Sect. 2. Then we introduce our concept of conducting model-checking within an agent during run-time in Sect. 3. Consequently we evaluate the concept on a lab plant by introduction of a new work piece (WP) type during operation. Finally we formulate a conclusion about the outcomes and an outlook about future research intentions.

2 State of the Art

2.1 Model-Checking

The idea of formal verification was originally based on finding a deductive mathematical proof of a software construct in regard to a specification formulated with temporal logic [1]. However theorem proofing is done manually and not applicable on large complex software programs, as implemented in programmable logic controllers (PLC) of aPS today. Therefore model-checking as an automatic and algorithmic search method has been introduced by Clarke, Emerson and Sifikais in the early 1980s. Since the 1990s model checking is also subject in automation science to enable formal verification for aPS.

Algorithmic approaches to verify hybrid systems containing discrete and continuous information about the aPS were developed by Silva et al. [2] and Stursburg et al. [3]. Hence, complex continuous models, which possess an infinite number of states, are firstly approximated by many small sub-models and subsequently discretized. Further approaches focused on model checking employing a probabilistic model for abstraction of networked automated systems were developed by Greifeneder and Frey [4] and Greifeneder et al. [5]. A comparison of the strengths and weaknesses of simulative and formal methods for the analysis of response time was also examined [6]. As conclusion, both methods have their own specific characteristics and are not equally well-suitable depending upon the aim of the analysis.

Focusing on verification of the PLC program code of aPS without considering the plant behaviour as part of the system was examined by Schlich and Kowalewski [7], Biallas et al. [8] and Kowalewski et al. [9] based on over-approximation. Further Kowalewski et al. implemented two optimizations to deal with the state space explosion problem [8]. Expanding this approach by integrating the aPS itself, or separating the whole control system into small sub-systems (e.g. agents) was not considered yet.

The necessity of integrating the plant model for model-checking of logic controllers focused on untimed properties was presented by Santiago and Faure [10] and Machado et al. [11]. Therein, the verification results obtained by the lack of a plant model, by usage of a simple plant model or usage of a detailed model, were compared. Faure et al. showed that only a detailed model of the plant is able to verify every defined property, by combining simulative and formal methods, to enable model-checking approaches to verify even complex aPS with continuous behaviour.

2.2 Agent-Based Software Engineering in Industrial Automation

One approach to modularize and distribute control in industrial automation is surveyed with the usage of intelligent agents, building a multi-agent system (MAS) [12, 13]. The definition of such agents and its communication is often based on the FIPA[1] standard for physical agents [14]. The current research on agents is largely conducted in manufacturing automation (cp. survey [15]) as well as in smart grid applications. These developments are concentrating to increase flexibility, efficiency (e.g. energy consumption) and fault-tolerance during operation execution

---

[1]Standards for interoperability among software agent platforms (FIPA), http://www.fipa.org, retrieved on 8/17/2015.

[16]. The design paradigm of MAS in industrial control assigns an agent a specific perception (e.g. the sensor data) and an associated action space in which it carries out tasks and manipulates the modules actuation.

Commonly the MAS is divided into a resource agent for a production module and a product agent for a WP. The decisions in the MAS lead to solving a given production request by allocating the tasks to the agent classes and therefore to dynamically generate a production schedule during run-time. The description of the capabilities and the technical process for manufacturing the WP is stored within the agent's knowledge-base and the exchange with other agents is conducted by employing a common ontology to enable message encoding/decoding. Decision making of an agent can be divided into two steps: deliberation and reasoning for deriving a plan for future execution [17]. Above the sheer possibility to offer a production step, it is not ensured that the action may be carried out without failure. However this is undertaken by reasoning and further poses a crucial step towards estimating the processing of a new WP type.

3   Concept for Agent-Based Model-Checking During Run-Time

To enable an aPS to produce a WP, which was not considered during the design, a concept for model checking at run-time based on MAS is presented. Hence, aPS is partitioned into modules, e.g. a handling or processing station, connected by logistic connections, e.g. belt drives. Regarding its flexibility, it is assumed that the WP's specifications are varying in parameters such as dimensions or weight, but not in its magnitude and general structure. In the scenario of a cyber-physical production system (CPPS), the request for a new product has to be considered for production in more than one plant and consequently the decision must take performance and safety indicators into account. In our case we realize this control by the multi agent paradigm. Two different kinds of agents are incorporated in our approach: first the *product agent*, generating a request for a new product in the form of a specification; secondly the *resource agent* in form of the entity, carrying out the production process. A redundant existence of functionalities in form of resources opens the solution space for a flexible behaviour of the plant.

In case of an unknown product request, a module must ensure safety and feasibility for production process execution. Consequently the agents need to carry a knowledge base about their environment and the modules respective their physical behaviour.

The model-checking mechanism is triggered after a deliberation process that determines which resource agents are compatible with the certain product steps of the product. The transformation of control software, e.g. PLC code, and semiconductor hardware architecture into a representation for model-checking is not focused here, since advances have already been conducted in this field in the last decade [18–20]. More importantly the behaviour of the plant must be derived in

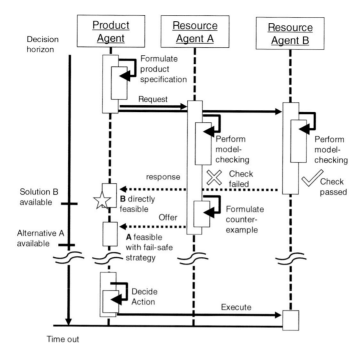

Fig. 1 Sequence diagram for model-checking triggered by a new product specification

form of a discrete state-model, e.g. a finite state-machine (FSM), by a manual modelling process.

In the current state-of-the-art it has been shown that continuous models (e.g. state-space, derived by ODE's) may be discretized automatically, but the burden of choosing a feasible discretization step while obtaining correct behaviour has to be resolved. In our approach no automatic discretization step is chosen, rather a FSM with parameterized and time-dependent state transitions is derived and automatically trained by a physical and continuous model. The states of the model represent an abstraction to discrete physical conditions of the component, while reducing the number of states by neglecting transitional states that have no influence on a certain failure mode in the system.

The model-checking process executes an algorithmic exhaustive search through the state-space of the concurrent control and plant model. Both are described with discrete states and conditional behaviour. If the concurrent execution of the models contradicts the specification, the model checker returns the respective state-combination as a counter-example. Generally the counter-example supports the search for a cause in the models faulty behaviour (e.g. state-transitions). Figure 1 presents a sequence diagram showing two resource agents which offer a production step. One conducts to switch the control to a fail-safe mechanism and restart the model-checking. After it would return without error, the production request can be offered to the product agent with an alternative offer.

4   Evaluation

For evaluation of the concept, a Pick and Place Unit (PPU) for handling cylindrical work pieces of different colours (white, black) and materials (plastic, metal) is used. Considering typical evolution scenarios of aPS', separated into sequential and parallel evolution, 16 scenarios based on the PPU were developed [21]. In the last evolution scenario the PPU contains the mechatronic modules separator, conveyor and stamp. For transportation of the WPs, an electric driven crane with a vacuum gripper is located between these modules.

To maximize the throughput of WPs transported by the crane, the acceleration and deceleration distances of the crane have to be optimized as short as possible, which increases the inert force on the WPs. Consequently WPs are oscillating shortly after the crane has stopped at the target area. Hence, based on the oscillation, WPs may be dropped beside the storage area, if the crane moves down directly after it has stopped. A sectional drawing of an oscillating WP at the storage area using an arbitrary set of parameters is shown in Fig. 2. Therein the y-axis is separated into three parts: around zero, positive and negative, which describes the position of the WP being above, right or left of the storage area, respectively.

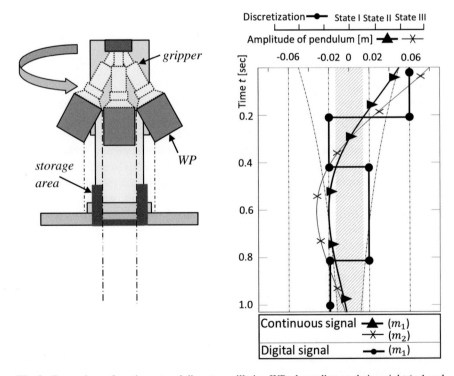

Fig. 2 Comparison of continuous and discrete oscillating WPs depending on their weight (m1 and m2) with an arbitrary set of parameters

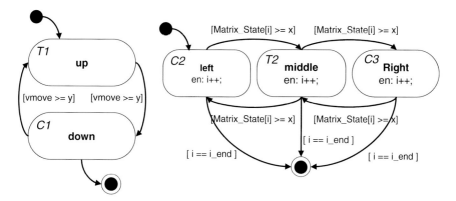

Fig. 3 Finite state machines for vertical (*left*) and horizontal (*right*) movement processes with clocks x and y

To approximate this complex and continuous physical mechanism containing infinite positions of the WP, an approach to describe the position of the WP using only three states was developed: directly over the storage area ("middle"), right of the storage area ("right"), and left of the storage area ("left") (cp. Fig. 3).

Intending to drop down the WP on the storage area, it has to be in state "middle" once the crane is retracted. Considering Fig. 2, there is only one short time range from 0.2 to 0.41 s, where this condition is met while the WP oscillates. In this example case, there is no maximization in throughput, WPs can be dropped down safely after 0.8 s, since the amplitude of the oscillation is too low to leave state I at anytime in the future.

After the transformation of the continuous model into two concurrent finite state machines, containing three states for horizontal and two for vertical movement, verification via model checking was applied.

Hence, the approximated model considered in Fig. 3 was implemented into the tool MATLAB/Simulink[2] Stateflow.

To verify whether the crane dropped down the WP correctly, the model needs two parameters describing the high and the retracting speed of the crane as well as the durations of the horizontal states. These values are pre-calculated by the continuous model running prior to Stateflow and are stored in the 1xk array Matrix_State. Depending on the parameter values and the number k of elements in the array, the approximated model returns the position of the WP, i.e. right, left or on the storage area. The number of transitions which are performed to determine the position of the WP are equal to the number of columns k of the array.

---

[2]http://www.mathworks.com, retrieved on 2/19/2016.

The resulting model checking problem can be compared to a proof for mutual exclusion. The problem formulated in computation tree logic (CTL) is written as EF(f) with f = (C1 && C2) ‖ (C1 && C3). This describes, that the model checker should search if EF(f) holds for the processes of vertical and horizontal movement to not enter the critical process C1–3 at the same time, meaning not to clamp or to drop the WP besides the drop zone. Considering the oscillation frequency of the crane, which does not depend on the mass of the WP, the verification results are also valid for WP, which mass is comparable to the verified mass (see Fig. 3 cross marked line). To transport WPs containing a high difference mass by the crane, further verification runs including pre-calculations by the continuous model are essential. Based on the low number of states, which are necessary to verify this mechanical phenomenon, run-time verifications in control systems, i.e. agents, are feasible.

5   Conclusion and Outlook

In this paper we demonstrated a concept to perform model-checking of an automation module during run-time, based on the integration into the resource agent of MAS. As requirement for this approach a model of the production resource by formulating its logical process steps into a discrete model (e.g. a FSM) has to be designed. To demonstrate the benefits and results, we applied the introduced agent-based approach to a pneumatic crane of a pick-and-place unit.

A demonstration on the real plant showed that a variation in product specification, i.e. mass, with standard control software resulted in oscillations of the product during handling. These oscillations resulted in clamping and therefore triggered a production fault. After applying the novel approach, the error was detected by the resource agent through model-checking and was communicated to the product agent a priori.

In the future we will focus our work on deriving models of the plant for classifying reasonable levels of abstractions of respective failure types. Further we will re-apply the method on all modules of the PPU and extend the process with redundant production resources to observe the behaviour with n model-checking agents in total. Further we will analyse the benefits of counter-examples and how to derive self-healing measures for the agent's action space. This would allow the agent to adapt the control logic during run-time to sustain production under varying product parameters and thus lead to a higher dependability of the complete aPS.

Acknowledgment We thank the German Research Foundation (DFG) for funding this project as part of the Priority Programme SPP 1593: Design for Future—Managed Software Evolution.

References

1. Clarke, E.M., Emerson, E.A., Sifakis, J.: Model checking: algorithmic verification and debugging. Commun. ACM **52**, 74 (2009)
2. Silva, B.I., Stursberg, O., Krogh, B.H., Engell, S.: An assessment of the current status of algorithmic approaches to the verification of hybrid systems. In: Proceedings of the 40th IEEE Conference on Decision Control (Cat. No.01CH37228), vol. 3, pp. 2867–2874 (2001)
3. Stursberg, O., Lohmann, S., Engell, S.: Improving dependability of logic controllers by algorithmic verification. In: IFAC World Congress 2005, pp. 104–109 (2005)
4. Greifeneder, J., Frey, G.: Probabilistic hybrid automata with variable step width applied to the anaylsis of networked automation systems. Discret. Syst. Des. **3**, 283–288 (2006)
5. Greifeneder, J., Liu, L., Frey, G.: Comparing simulative and formal methods for the analysis of response times in networked automation systems. In: IFAC World Congress 2008, vol. 1, p. O3 (2008)
6. Kwiatkowska, M., Norman, G., Parker, D.: PRISM: probabilistic symbolic model checker. Comput. Perform. Eval. Model. Tech. Tools **2324**, 200–204 (2002)
7. Schlich, B., Kowalewski, S.: Model checking C source code for embedded systems. Int. J. Softw. Tools Technol. Transf. **11**, 187–202 (2009)
8. Biallas, S., Kowalewski, S., Stattelmann, S., Schlich, B.: Efficient handling of states in abstract interpretation of industrial programmable logic controller code. In: Workshop on Discrete Event Systems (WODES 2014), pp. 12–17, Cachan, France (2014)
9. Kowalewski, S., Engell, S., Preußig, J., Stursberg, O.: Verification of logic controllers for continuous plants using timed condition/event-system models. Automatica **35**, 505–518 (1999)
10. Santiago, I.B., Faure, J.-M.: From fault tree analysis to model checking of logic controllers. In: IFAC World Congress 2005, pp. 86–91 (2005)
11. Machado, J.M., Denis, B., Lesage, J.J., Faure, J.M., Ferreira Da Silva, J.C.L.: Logic controllers dependability verification using a plant model. In: Discrete Event Systems, vol. 3, pp. 37–42 (2006)
12. Wooldridge, M., Jennings, N.: Intelligent Agents: Theory and Practice (1995)
13. Colombo, A., Schoop, R., Neubert, R.: An agent-based intelligent control platform for industrial holonic manufacturing systems. IEEE Trans. Ind. Electron. **53**, 322–337 (2006)
14. Poslad, S.: Specifying protocols for multi-agent systems interaction. In: ACM Trans. Auton. Adapt. Syst. **2**, 15 (2007)
15. Leitao, P., Marik, V., Vrba, P.: Past, present, and future of industrial agent applications. IEEE Trans. Ind. Inf. **9**, 2360–2372 (2013)
16. Schütz, D., Wannagat, A., Legat, C., Vogel-Heuser, B.: Development of plc-based software for increasing the dependability of production automation systems. IEEE Trans. Ind. Inf. **9**, 2397–2406 (2013)
17. Wooldridge, M.: An Introduction to Multiagent Systems, 2nd edn. Wiley, New York (2009)
18. Younis, M.B., Frey, G.: Formalization of existing plc programs: a survey. In: CESA 2003, Lille (France), Paper no. S2-R. -00–0239 (2003)
19. Schlich, B., Brauer, J., Wernerus, J., Kowalewski, S.: Direct model checking of PLC programs in IL. In: 2nd IFAC Workshop on Dependable Control of Discrete Systems, pp. 28–33 (2009)
20. Schlich, B.: Model Checking of Software for Microcontrollers (2008)
21. Vogel-Heuser, B., Legat, C., Folmer, J., Feldmann, S.: Researching evolution in industrial plant automation: scenarios and documentation of the pick and place unit (2014)

A Synchronous CNP-Based Coordination Mechanism for Holonic Manufacturing Systems

Doru Panescu and Carlos Pascal

Abstract Our paper presents a new holonic coordination approach, which can be useful for the difficult case when more holons try to assign common resources. It is based on the Contract Net Protocol and this is combined with synchronous back-tracking. The coordination scheme also implies an a priori established hierarchy of manager holons. The proposed method ensures a safe operation and was evaluated on a manufacturing scenario with four robots.

Keywords CNP · Holonic system · Coordination · Robotized system

1 Introduction

Holonic manufacturing systems (HMSs) belong to architectures that can be a step in the change from the centralized control to adaptable, decentralized approaches [1, 2]. HMSs still have a reduced deployment, to be explained through the coordination mechanisms that can be unsafe. This paper presents the first stage of a research for obtaining sound holonic coordination. While maintaining the distributed part of a holonic scheme, it can be a right idea to organize coordination so that to avoid nondeterministic behaviours and wrong results. Our proposal is to keep the structure of holonic architectures—like in PROSA [3] or HAPBA [4], and do coordination in a synchronous, hierarchical manner. Based on our previous research [2], we take into account the Contract Net Protocol (CNP) [5], for the case close to reality, i.e. when more managers try to solve their goals in the same time, using common resources. An inspiration point was the Distributed Constraint Satisfaction Problem (DisCSP), a method that can be used for holons' coordination, too [6].

D. Panescu (✉) · C. Pascal
Department of Automatic Control and Applied Informatics, "Gheorghe Asachi" Technical University of Iasi, Iasi, Romania
e-mail: dorup@ac.tuiasi.ro

C. Pascal
e-mail: cpascal@ac.tuiasi.ro

© Springer International Publishing Switzerland 2016
T. Borangiu et al. (eds.), *Service Orientation in Holonic and Multi-Agent Manufacturing*, Studies in Computational Intelligence 640,
DOI 10.1007/978-3-319-30337-6_16

Literature shows more adaptations of CNP [2, 7, 8]; even so, to the best of our knowledge there is no scheme to use a synchronous, hierarchical coordination of agents.

2   A Synchronous Holonic Coordination Scheme

2.1   Hypotheses of the Proposed Coordination Protocol

Before detailing the new coordination scheme, the used assumptions are presented.

- If the HMS operation involves a single manager holon, then the system functions according to the normal CNP.
- When more managers operate, there is an a priori established hierarchy between them; this means there is a manager having the most important tasks to solve, which consequently has the highest priority, and so on, until the less important manager that has the lowest priority. This hierarchy is decided at a certain instant, supposing that all manufacturing commands are known at that time. Orders that are received during negotiation will be considered in the next deliberation phase. The way managers create holarchies for their goals' solving is a distributed one, according to the negotiation carried out with resource holons.
- Each manager knows which is its successor and predecessor, respectively.
- Any manager (an order holon or product holon) may have to solve one or more tasks; these are goals for resource holons, which have the role of contractors in CNP. The pairing between managers' goals and contractors' is so that the case when a contractor becomes manager is excluded (it is supposed that any goal is to be solved by a single contractor). Thus, the set of managers is a priori known and not modified during the coordination process. In what follows we call a manager's assignment the *pairing* established between its goals and contractors.
- Managers communicate among themselves with two types of messages as in DisCSP; these messages are adapted to the specific of CNP. We use Ok and Ng_Ok messages. An Ok message is sent from a manager to its successor to announce the contractors being used by itself and its predecessors. Each manager appends or updates the received Ok message with its own assignment. As Ok message are used to also announce the result after a backtracking process, there are two types of Ok messages: positive (labelled Ok+) and negative (Ok−). A Ng_Ok message is issued when an agent cannot solve a goal according to the bids received from contractors. This is sent from an agent to its predecessor. The Ng_Ok messages contain two parts: a Nogood part which includes information on contractors that are needed by a manager to satisfy its goals and are requested from higher priority agents, and an Ok part which contains the updated situation on the contractors used by managers. This part is continuously updated by managers, as explained below. It is to be understood that a Ng_Ok message

launches a backtracking process. Depending on how the requests included in the Ng_Ok message could be satisfied or not, the search process continues with an Ok+ or Ok− message. The last manager, with the lowest priority, detects the end of coordination.

• We consider the protocol adjusted so that a manager takes a rough decision: it either can solve all its goals and thus it continues with the decided assignment, or if there is at least one goal that it cannot solve, then it renounces to all its goals and makes no assignment (it engages no contractor).

There are three cases regarding the bidding process: a positive bid is sent to a manager when a contractor can solve a goal and it is free (not committed to another manager); a negative bid is sent when though the contractor can solve the goal, it is already committed to another manager; no bid is sent when the contractor cannot solve the goal (it does not have the needed capacity). About this, it is supposed that bids can be sent from contractors to managers in a specified time, so that a manager knows which is the interval needed to wait for bids. In the next section, only the operation of managers is commented, because for contractors this is the same as in the normal CNP, except for contractors' sending negative bids, which is not usual in CNP.

2.2 Operation of Managers with Different Positions in Hierarchy

In what it follows agents having the role of managers in CNP are labelled with M, while contractors being resource holons are labelled with R. Due to the synchronous operation of the proposed coordination scheme, in which at any time a single manager is active, we have to describe the operation of the first manager (the one with the highest priority), of an intermediary manager and of the last manager.

(a) Operation of manager M_1

```
Apply CNP
if all goals are solved
  then Send Ok+ to M₂
  else Send Ok+ to M₂ with no assignment

when Ng_Ok is received do
if Solve(Ng_Ok)is Positive
  then Send Ok+ to M₂        else Send Ok- to M₂
```

Manager M_1 starts the search process. At that time, it is supposed that all managers have already received the input data so that they know their goals. M_1 applies CNP. This regards broadcasting the goals to all contractors, the bidding process, the decision on chosen bids and the announcement of selected contractors

[5]; the execution part of CNP is not discussed in this paper. If the result of CNP is positive (all goals could be solved), then M_1 sends an Ok+ message to M_2, which is: (Ok+, (M_1, C_{M1})), where C_{M1} is the list of contractors that were chosen by M_1. According to one of the hypotheses presented in the previous section, if M_1 cannot satisfy all its goals, then it sends an Ok+ message with no assignment, namely: (Ok +, $(M_1, ())$). Being the agent with the highest priority, M_1 can receive a Ng_Ok message from its successor. In this case, it applies a procedure to solve the request (this is explained latter); if the result is positive, it continues with an Ok+ message, otherwise with an Ok− message.

(b) Operation of manager M_i

```
when Ok+ is received do
  if no assignment was made or a Ng_Ok was initiated
     then Apply CNP
        if all goals are solved
           then Send Ok+ to M_{i+1}
           else if there is at least one goal with no bid
                   then Send Ok+ to M_{i+1} with no assignment
                   else Send Ng_Ok to M_{i-1}
     else Send Ok+ to M_{i+1}
when Ok- is received do
  if a Ng_Ok was initiated
     then Send Ok+ to M_{i+1} with no assignment
     else Send Ok- to M_{i+1}
when Ng_Ok is received do
  if Solve(Ng_Ok) is Positive
     then Send Ok+ to M_{i+1}   else Send Ng_Ok to M_{i-1}

Solve(Ng_Ok)
  if all requested resources are freed by the set of bids
     then Change assignment to free requested resources
          Update Ok
          return Positive
     else Change assignment as much as possible
          Update Ng_Ok
          return Negative
```

The operation of an intermediary agent starts when it receives a first Ok+ message from its predecessor; by that time, it has not yet made any assignment. A first difference with respect to M_1 is for the case when manager M_i has one or more goals for which it received only negative bids. In this case, it composes a Ng_Ok message, with its two parts as specified in the previous section. As an example, let us suppose that M_i had three goals to solve ($G_1 \div G_3$), for which it received the following bids: $G_1 \leftarrow R_1+, R_2-; G_2 \leftarrow R_3-; G_3 \leftarrow R_4-, R_5-$ (for G_1 it received a positive bid from contractor R_1 and a negative one from R_2, and so on).

In this case the Ng_Ok message composed by M_i is: (Nogood; (M_i, ((R_3), (R_4, R_5))); Ok; ((M_1, C_{M1}), ..., (M_i, (R_1)))), by which it requests the releasing of resource R_3 and R_4 or R_5. Then it waits for receiving a consequent Ok message; if this is an Ok+ it means that the requested contractors could be freed by its predecessors and a right assignment will be found after applying the CNP. If it receives an Ok− message, its request could not be satisfied and thus M_i continues with an Ok+ sent to M_{i+1}, indicating no assignment made by itself.

Being in intermediary manager, M_i can receive a Ng_Ok message from a successor. In this case it applies the procedure Solve which uses the data of the received Ng_Ok message. Each manager, after applying CNP, keeps the whole list of received bids. Using this list, the manager can determine whether it can release all the requested resources—in this case the procedure Solve returns Positive, or not when Solve returns Negative. In the first case M_i must continue with a correspondingly updated Ok+ message to be sent to M_{i+1}. Otherwise, it updates the Ng_Ok message if it can release part of the requested contractors and sends this message to its predecessor, so that backtracking is continued. As an example, let us suppose that manager M_{i-1} received the above considered Ng_Ok message and it can release only the resource R_4. Namely, according to the received bids, it happens that it can use R_6 instead of R_4. In this case the Ng_Ok message that will be sent by M_{i-1} to M_{i-2} is: (Nogood;(M_i, ((R_3))); Ok; ((M_1, C_{M1}), ..., (M_{i-1}, ... R_6 ...), (M_i, (R_1)))).

A further case is when an intermediary agent receives an Ok message after a backtracking process was initiated. Here we have two situations. If the manager is not the one that initiated the backtracking process, then it only has to pass the message unchanged to its successor. If the manager is the one that initiated a Ng_Ok message, then depending on the type of received message (Ok+ or Ok−), it will be able to make an assignment for all its goals or it fails. In both cases, it sends an Ok+ message to its successor, so that the search should be continued.

(c) Operation of manager M_n

```
when Ok+ is received do
  Apply CNP
  if all goals are solved
    then Broadcast Ok+ and End_of_search
    else Send Ng_Ok to M_{n-1}
when Ok- is received do
  Change assignment freeing resources
  Broadcast Ok+ and End_of_search
```

M_n can receive only Ok messages. When it receives an Ok+ message, it can either find a solution for its goals and then compose the ultimate Ok message including its assignment and broadcast the end of search to all agents, or it issues a Ng_Ok message. Depending on the received result (an Ok+ or Ok− message), it will correspondingly end the coordination process.

2.3   An Illustrative Case Study

To show the utility of the above presented mechanism, a manufacturing problem is considered and solved according to the proposed coordination scheme. The case study is inspired by the experimental system existing in our Robotics laboratory, which is shown in Fig. 1 [9]. The manufacturing goals for which we can make experiments regard assembling/palletizing tasks solved by a robotized system with two industrial robots, which have a common workspace. The example presented in Fig. 1 is for a case when the two robots must fill in a pallet with certain parts. A more complicated experiment is taken into account in this paper in order to better demonstrate the operation of the proposed method. Namely, as displayed in Fig. 2, the environment contains four robots (labelled $R_1 \div R_4$) that operate in a common area, having to fill in four pallets ($P_1 \div P_4$) with different parts. Near each robot there is a storage containing the parts which can be used by that robot. According to its position, each robot has access to only two pallets, as follows: robot R_1 operates with P_1 and P_3, R_2 with P_1 and P_2, R_3 with P_2 and P_4, and R_4 with P_3 and P_4. The manufacturing goals regard the content of the four pallets. The corresponding HMS is one with four product holons representing the pallets with their needed content, while robots become resource holons. The most important goal is for pallet P_1 and

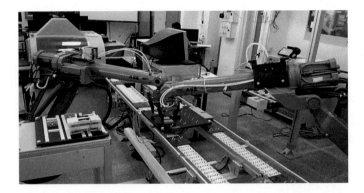

Fig. 1 The manufacturing environment that inspired the case study used in this paper

Fig. 2 The two instances of the analysed manufacturing problem

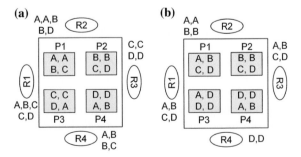

thus the corresponding product holon gets the highest priority. The following priorities are for holons representing pallets P_2, P_3 and P_4.

The first instance of our problem is for the initial state displayed in Fig. 2a. The parts are of four types: A, B, C and D. The storages of robots contain parts as shown in Fig. 2a near each robot, and the needed content of pallets is presented, too. The coordination process is started by M_1. For its four goals (a goal is issued by M_1 for each part to be placed in P_1) it receives bids from robot resource holons R_1 and R_2. The received bids (according to the contents of robots' storages) are: R_1 + A; R_1 + B; R_1 + C; R_2 + A; R_2 + A; R_2 + B. Bids are marked with the name of robot and name of part to be transferred in P_1, and all bids are positive because these are the first bids made by contractors. From this set of bids, M_1 selects a solution which is communicated to contractors (each contractor will know whether it was selected by M_1 or not, so that to make corresponding bids for next managers) and is communicated through an Ok+ message to manager M_2. Namely, let us suppose that M_1 decided the solution corresponding to the message: (Ok+; ((M_1, (R_1 + B, R_1 + C, R_2 + A; R_2 + A)))).

After receiving this Ok message, M_2 applies CNP for its goals and receives the following bids: R_2 + B; R_2 + B; R_2 + D; R_3 + C; R_3 + D. M_2 selects a combination of bids that satisfies all its goals and communicates the corresponding Ok+ message to M_3: (Ok+; ((M_1, (R_1 + B, R_1 + C, R_2 + A, R_2 + A)), (M_2, (R_2 + B, R_2 + B, R_3 + C, R_3 + D)))). Then, M_3 applies CNP to solve its goals and receives the bids: R_1 + A; R_1 + C; R1 + D; R_1 − C; R_4 + A; R_4 + C. From this set of bids M_3 can find a solution and correspondingly sends the following Ok+ message to M_4: (Ok+; ((M_1, (R_1 + B, R_1 + C, R_2 + A, R_2 + A)), (M_2, (R_2 + B, R_2 + B, R_3 + C, R_3 + D)), (M_3, (R_1 + C, R_1 + D, R_4 + A, R_4 + C)))). When the last agent applies CNP, it receives the following bids: R_3 + D; R_3 − D; R_4 − A; R_4 + B. From this set of bids M_4 cannot get a solution for all its goals; namely it did not receive any positive bid for a part A and a part D. So it starts backtracking by issuing a Ng_Ok message to its predecessor: (Nogood; (M_4, (R_3 − D, R_4 − A)); Ok; ((M_1, (R_1 + B, R_1 + C, R_2 + A, R_2 + A)), (M_2, (R_2 + B, R_2 + B, R_3 + C, R_3 + D)), (M_3, (R_1 + C, R_1 + D, R_4 + A, R_4 + C)) (M_4, (R_3 + D, R_4 + B)))). It is to notice that the Ok part of message included the contractors reserved by M_4 in its attempt to solve the goals. M_3 receives the above message and tries to solve it without jeopardizing the solution for its goals. Thus, it happens that M_3 can free the contractor R_4 with the part A, by choosing another bid for that goal, namely the one of contractor R_1. The other request of the Ng_Ok message (R_3 with part D) cannot be solved by M_3 because in fact it did not engage contractor R_3. Thus, M_3 creates a new Ng_Ok message with updated Nogood and Ok parts, which is sent to M_2: (Nogood; (M_4, (R_3 − D)); Ok; ((M_1, (R_1 + B, R_1 + C, R_2 + A, R_2 + A)), (M_2, (R_2 + B, R_2 + B, R_3 + C, R_3 + D)), (M_3, (R_1 + C, R_1 + D, R_1 + A, R_4 + C)), (M_4, (R_3 + D, R_4 + B)))).

According to the previously received set of bids, M_2 discovers that it can free the contractor R_3 with the part D, as it can use the bid for part D from R_2. Thus it creates an Ok+ message (as there is nothing more to be solved in the Ng_Ok message) with its part correspondingly up-dated. This is: (Ok; (M_1, (R_1 + B, R_1 + C, R_2 + A, R_2 + A))(M_2, (R_2 + B, R_2 + B, R_2 + D, R_3 + C))(M_3, (R_1 + C,

Table 1 Coordination process for the second instance of the manufacturing problem

Step	Agent	Received bids	Issued message
1	M_1	$R_1 + A; R_1 + B; R_1 + C;$ $R_1 + D; R_2 + A; R_2 + B$	$(Ok+; ((M_1, (R_1 + B, R_1 + C, R_1 + D;$ $R_2 + A))))$
2	M_2	$R_2 + B; R_2 + B; R_3 + B;$ $R_3 + C; R_3 + D$	$(Ok+; ((M_1, (R_1 + B, R_1 + C, R_1 + D, R_2 + A)),$ $(M_2, (R_2 + B, R_2 + B, R_3 + C, R_3 + D))))$
3	M_3	$R_1 + A; R_1–D; R_4 + D;$ $R_4 + D$	$(Nogood; (M_3, (R_1 - D)); Ok; ((M_1, (R_1 + B,$ $R_1 + C, R_1 + D, R_2 + A)), (M_2, (R_2 + B,$ $R_2 + B, R_3 + C, R_3 + D)), (M_3, (R_1 + A,$ $R_4 + D; R_4 + D))))$
4	M_2	Same previous set	$(Nogood; (M_3, (R_1 - D)); Ok; ((M_1, (R_1 + B,$ $R_1 + C, R_1 + D, R_2 + A)), (M_2, (R_2 + B,$ $R_2 + B, R_3 + C, R_3 + D)), (M_3, (R_1 + A,$ $R_4 + D; R_4 + D))))$
5	M_1	Same previous set	$(Ok-; ((M_1, (R_1 + B, R_1 + C, R_1 + D, R_2 + A)),$ $(M_2, (R_2 + B, R_2 + B, R_3 + C, R_3 + D)), (M_3,$ $(R_1 + A, R_4 + D; R_4 + D))))$
6	M_2	Same previous set	$(Ok-; ((M_1, (R_1 + B, R_1 + C, R_1 + D, R_2 + A)),$ $(M_2, (R_2 + B, R_2 + B, R_3 + C, R_3 + D)), (M_3,$ $(R_1 + A, R_4 + D; R_4 + D))))$
7	M_3	Same previous set	$(Ok+; ((M_1, (R_1 + B, R_1 + C, R_1 + D, R_2 + A)),$ $(M_2, (R_2 + B, R_2 + B, R_3 + C, R_3 + D), (M_3,$ $())))$
8	M_4	$R_3 + A; R_3 + B; R_3 - D;$ $R_4 + D; R_4 + D$	$(Ok+; ((M_1, (R_1 + B, R_1 + C, R_1 + D, R_2 + A)),$ $(M_2, (R_2 + B, R_2 + B, R_3 + C, R_3 + D), (M_3, ())$ $(M_4, (R_3 + A; R_3 + B; R_4 + D; R_4 + D))))$

$R_1 + D, R_1 + A, R_4 + C)) (M_4, (R_3 + D, R_4 + B))$. This message arrives to M4. It will re-apply the CNP mechanism and will find a solution for its goals. Then, it announces the end of searching process.

The second instance of the manufacturing problem is for the case of Fig. 2b. The evolution of coordination process is presented in Table 1. This time the back-tracking process launched by M_3 terminates unsuccessfully. Even so, its successor (M_4) can solve all its goals, and thus resource assignment is successfully ended for three of the four managers, which is the correct result according to the existing parts.

3   Conclusion and Future Work

This paper proposes a holonic coordination mechanism for cases that can frequently appear in practice, namely when more entities are competing for the same resources. The merit of the introduced coordination scheme is that it always determines a solution when there exists one. This happens because at any time only one manager operates with contractors (it means no blocking between managers is possible), and for each manager an exhaustive search is made (through back-tracking). Thus, a solution for a manager is not found only when either the manufacturing environment does not have the needed resources, or if these exist they are used by managers with higher priorities. Thus, an HMS or in general a multiagent system operating according to the proposed scheme can guarantee solutions for goals, in the order of their importance. In comparison with our previous approaches [2, 8], the present one has the advantage of converging to the optimal result. As a weak point, our mechanism is time consuming, due to the backtracking process and the way managers operate successively. For certain cases of manufacturing planning processes, time may not be critical, and thus the method should be applicable. As future work, we want to make further tests on virtual and real environments, complex problems and to see if it is possible to apply a distributed asynchronous backtracking approach, while keeping the safe operation.

References

1. Vrba, P., Tichy, P., Marik, V., Hall, K., Staron, R., Maturana, F., Kadera, P.: Rockwell automation's holonic and multiagent control systems compendium. IEEE Trans. Syst. Man Cybern.—Part C: Appl. Rev. **41**(1), 14–30 (2011)
2. Panescu, D., Pascal, C.: An extended contract net protocol with direct negotiation of managers. In: Borangiu, T., Trentesaux, D., Thomas, A., (eds.) Service Orientation in Holonic and Multi Agent Manufacturing and Robotics. Studies in Computational Intelligence, vol. 544, pp. 81–95. Springer, Berlin (2014)
3. Van Brussel, H., Wyns, J., Valckenaers, P., Bongaerts, L., Peeters, P.: Reference architecture for holonic manufacturing systems: PROSA. Comput. Ind. **37**, 255–274 (1998)
4. Panescu, D., Pascal, C.: On a holonic adaptive plan-based architecture: planning scheme and holons' life periods. Int. J. Adv. Manuf. Tech. **63**(5–8), 753–769 (2012)
5. Smith, R.G.: The contract net protocol: high level communication and control in a distributed problem solver. IEEE Trans. Comput. **C-29**, 1104–1113 (1980)
6. Pascal, C., Panescu, D.: A petri net model for constraint satisfaction application in holonic systems. In: IEEE International Conference on Automation, Quality and Testing, Robotics (AQTR), Cluj-Napoca, Romania (2014)
7. Kim, H.M., Wei, W., Kinoshita, T.: A new modified CNP for autonomous microgrid operation based on multiagent system. J. Electr. Eng. Technol. **6**(1), 139–146 (2011)
8. Panescu, D., Pascal, C.: On the staff holon operation in a holonic manufacturing system architecture. In: Proceedings of ICSTCC, Sinaia, Romania, pp. 427–432 (2012)
9. Panescu, D., Pascal, C.: HAPBA—a holonic adaptive plan-based architecture. In: Borangiu, T., Thomas, A., Trentesaux, D. (eds.) Service Orientation in Holonic and Multi-Agent Manufacturing Control. Studies in Computational Intelligence, pp. 61–74. Springer, Berlin (2012)

Interfacing Belief-Desire-Intention Agent Systems with Geometric Reasoning for Robotics and Manufacturing

Lavindra de Silva, Felipe Meneguzzi, David Sanderson, Jack C. Chaplin, Otto J. Bakker, Nikolas Antzoulatos and Svetan Ratchev

Abstract Unifying the symbolic and geometric representations and algorithms used in AI and robotics is an important challenge for both fields. We take a small step in this direction by presenting an interface between geometric reasoning and a popular class of agent systems, in a way that uses some of the agent's available constructs and semantics. We then describe how certain kinds of information can be extracted from the geometric model of the world and used in agent reasoning. We motivate our concepts and algorithms within the context of a real-world production system.

Keywords BDI agents · Geometric reasoning · Robotics · Manufacturing system

L. de Silva (✉) · D. Sanderson · J.C. Chaplin · O.J. Bakker
N. Antzoulatos · S. Ratchev
Faculty of Engineering, Institute for Advanced Manufacturing,
University of Nottingham, Nottingham, UK
e-mail: lavindra.silva@nottingham.ac.uk

D. Sanderson
e-mail: david.sanderson@nottingham.ac.uk

J.C. Chaplin
e-mail: jack.chaplin@nottingham.ac.uk

O.J. Bakker
e-mail: otto.bakker@nottingham.ac.uk

N. Antzoulatos
e-mail: nikolas.antzoulatos@nottingham.ac.uk

S. Ratchev
e-mail: svetan.ratchev@nottingham.ac.uk

F. Meneguzzi
Pontifical Catholic University of Rio Grande Do Sul, Porto Alegre, RS, Brazil
e-mail: felipe.meneguzzi@pucrs.br

© Springer International Publishing Switzerland 2016   179
T. Borangiu et al. (eds.), *Service Orientation in Holonic and Multi-Agent Manufacturing*, Studies in Computational Intelligence 640,
DOI 10.1007/978-3-319-30337-6_17

1 Introduction

Modern manufacturing environments require systems capable of dynamically adjusting to rapid changes in throughput, available production equipment, and end-product specifications. When there are complex and non-specialized machines or robots involved able to perform a multitude of tasks, intelligent and flexible systems are needed for modelling parts, the environment and production processes, and for reasoning about how processes should manipulate parts in order to obtain the desired product. These systems typically reason in terms of concepts such as a machine's degrees of freedom, the positions and orientations of parts, and collision-free trajectories when moving parts during production. Such geometric reasoning is especially appealing in the context of manufacturing because detailed CAD models of parts and end-products are readily available, and production processes are often well defined. When a production system is controlled by a higher level software entity such as an agent system, typically reasoning in terms of abstract and symbolic representations that ignore the finer details present in the geometric model, it is crucial to unify at least some aspects of the two representations so that they may be linked and information shared. Indeed, such a unified representation is an important challenge for robotics and AI in general.

This paper focuses on interfacing a (single-agent) agent programming language, from the popular Belief-Desire-Intention (BDI) family of agents [1], with geometric reasoning in a way that exploits some of the agent's existing constructs and semantics. We also give insights into the information types that can be abstracted from the geometric model for the agent's benefit; this includes information about any new, a priori unknown objects in the domain, and which objects are interconnected and will therefore move together. Since BDI agent systems do not plan their actions before execution, but instead perform context-based expansion of predefined (user-supplied) plans during execution, our work differs from existing works such as [2–8] which focus on integrating symbolic *planners* with geometric reasoning entities. A notable exception is [9], who also interleaves symbolic reasoning with acting as we do; however, this work does not use a standard model of agency.

2 Background

Geometric Reasoning. In this paper we use the term *geometric reasoning* to refer to motion planning as defined in [10]. A *state*, then, is the 3D world $W = \mathbb{R}^3$, and its fixed obstacles are the subset $O \subset \mathbb{R}^3$. A robot is modelled as a collection of (possibly attached) rigid *bodies*. For example, a simple polygonal robot A could be defined as the sequence $A = (x_1, y_1, z_1), \ldots, (x_n, y_n, z_n)$, where each $(x_i, y_i, z_i) \in \mathbb{R}^3$. A key component of motion planning is a *configuration space* C which defines all the possible transformations that can be applied to a body such as A above. More specifically, a *pose* (or *configuration*) $c \in C$ is the tuple $c = \langle x, y, z, h \rangle$, where $(x, y, z) \in \mathbb{R}^3$ and h is the unit quaternion, i.e. a four dimensional vector used to perform 3D rotations; in a 2D world $W = \mathbb{R}^2$, h would instead be an angle in $[0, 2\pi)$. With a slight abuse of notation the transformation of a body A by pose c is denoted $A(c)$. A robot's pose composed of bodies $A_1, \ldots, A_n$ is an element of $C_1 \times \ldots \times C_n$, where each C_i is the configuration space of A_i. If a body A_2 is attached via a joint to the end of some body A_1, some of A_2's degrees of freedom will be constrained, e.g. the x, y and z parameters of all poses of A_2 might depend on the corresponding ones in A_1.

A motion planning problem, then, is a tuple C, col, c_I, c_G, where $col : C \rightarrow \{true, false\}$ is a function from poses to truth values indicating whether a pose $c \in C$ is in collision ($col(c) = true$) with some object or not, and $c_I, c_G \in C$ are the initial and goal poses [3]. A collision-free motion plan solving a motion planning problem is a sequence $\boldsymbol{c} = c_1, \ldots, c_n$ such that $c_I = c_1, c_G = c_n$, and for each pose c_i, we have $c_i \in C$ and $col(c_i) = false$.

BDI Agents. In this work we use the popular AgentSpeak [1] agent programming language to formally represent the large class of BDI agent systems in the literature. An AgentSpeak agent is a tuple $Ag = \langle E, B, Pl, I \rangle$ where: E, the event queue, is a set consisting of both external events (environment perception) and internal events (subgoals); B, the belief base, is a set of ground logical atoms; Pl, the plan library, is a set of plan-rules; and I, the intention stack, is a set of partially instantiated plan steps of plan-rules that were adopted. A **plan-rule** is a syntactic construct of the form $\langle e \rangle : \langle con \rangle \leftarrow \langle body \rangle$, where $\langle e \rangle$ is the triggering event; $\langle con \rangle$, the context condition, is a logical formula; and $\langle body \rangle$ is a sequence of steps to be executed in the environment. There are two types of triggering events relevant to this paper: $+\varphi$ or $-\varphi$ for an atom φ indicates, respectively, that a belief in B has been added or removed, and $-!\psi$ or $-?\psi$ indicates, respectively, that an achievement or test goal has failed, i.e. that either the plan to achieve ψ has failed during execution, or that belief ψ does not hold in B, respectively. Finally, $\langle body \rangle$ is constructed from the following elements: (i) the execution of an action in the environment; (ii) the adoption of a subgoal $!\psi$ or testing of a condition $?\psi$, both of which generate internal events; or (iii) the explicit modification of a belief ($+\varphi$ or $-\varphi$). An example of a plan-rule in our AgentSpeak-like language is[1]:
$+!mov(R, F, T) : canMov(R, F, T) \leftarrow nav(R, F, T); ?at(R, T).$

If the achievement goal $!mov(r1, t1, t2)$ is reached when some rule is executed, AgentSpeak looks up Pl for a rule that is both relevant and applicable for the goal. Our rule above is relevant because $mov(R, F, T)$ and $mov(r1, t1, t2)$ unify on the application of substitution $\theta = \{R/r1, F/t1, T/t2\}$ to the former; we use $mov(\boldsymbol{o})$ to denote the ground instance resulting from operation $mov(\boldsymbol{v})\theta$, where $\boldsymbol{v}$ and $\boldsymbol{o}$ are the vectors of variables and constants above. If the plan-rule is also applicable, i.e. belief $canMov(r1, t1, t2) \in B$, then the plan's body, after applying the substitution, is added to I as a new intention. Pursuing it involves executing action $nav(r1, t1, t2)$

---

[1]R is short for *Robot*, F for *From*, T for *To* and ti for *table i*.

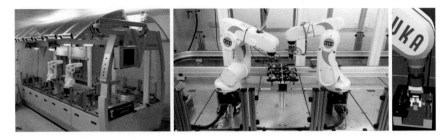

Fig. 1 The assembly platform, tool rack, and a simulation of the pallet being gripped

and then checking for success by testing B via ? $at(r1, t2)$. The action to navigate is defined by the following action-rule: $nav(R, F, T) : at(R, F) \wedge canMov(R, F, T) \leftarrow mvExec(R, F, T); mvEff()$, where $mvExec(R, F, T)$ is associated with a procedure that moves the robot, and $mvEff()$ with one that returns, possibly after sensing the environment, a set of literals representing the result of moving.[2]

The Assembly Platform. We use the production system in Fig. 1 [11] as a running example to motivate some of the concepts in this paper. The system combines the functionality of six independent workstations, each governed by a separate agent, to assemble detent hinges for lorry-cab furniture. Each station is served by a linear transfer system that transports a pallet carrier; this supports a pallet with the individual parts that need to be assembled, as well as the partially/fully assembled hinge. The six workstations, controlled by PLCs (Programmable Logic Controllers), are as follows: two consist of a Kuka robot each; two accommodate one workspace each; one contains a tool changing rack; and one contains an inspection station. The tool changing rack is placed between the Kuka arms, which have access to the rack as well as to the workspaces that are used for carrying out assembly operations. The rack contains six slots which can hold up to six different types of end effectors such as pneumatic and two-finger grippers. RFID tags on the tools are used to determine which of them are currently on the rack, so that the Kuka arms may dynamically lock into the relevant ones during assembly. Finally, the inspection station is used to perform force and vision tests to verify whether the hinge was assembled correctly. The hinge that is assembled is composed of two separate leaves held together by a metal pin. Three metal balls need to be placed into adjacent cylindrical slots in the center of the hinge, three springs need to be placed into the same slots, and a retainer is used to close the hinge. By using only a subset of these parts to assemble a hinge, there can be four product variants, each having a different detent force.

---

[2]An action-rule's body is adapted from STRIPS to be a sequence of functions that return a (possibly empty) set of literals, each of which is applied to the belief base B, i.e. the positive literals are added to B, and atoms associated with negative literals are removed from B.

3   Interfacing AgentSpeak with Geometric Reasoning

Like in works such as [2, 7], evaluable predicates are fundamental in linking AgentSpeak with geometric reasoning. While standard predicates are evaluated by looking up the agent's belief base, evaluable predicates are attached to external procedures, which for us involve searching for a viable trajectory within a geometric world/state W. Thus, we call such predicates *geometric predicates*. For example, predicate $canMov(R, curr, T)$ in our plan-rule from the previous section could be a geometric predicate which invokes a motion planner to check whether it is possible for Kuka arm R to move from its current pose $curr$ to tool T, specifically, to a position from where the arm can now easily lock into the tool with a predefined vertical motion. We use $curr$ as a special constant symbol to represent the current pose.

To evaluate a geometric predicate it needs to be associated with a collection of *goal poses*, from which at least one needs to have a viable trajectory from the current pose for the predicate to evaluate to *true*. Goal poses could either be determined manually or computed offline automatically with respect to the 3D model of the world and the objects involved. In our assembly platform, for example, the Kuka arms are manually trained on how to grasp the various shapes that might be encountered during production. This is especially important because objects like the pallet carrier are too heavy to be lifted from most seemingly good grasps and poses—there is only one pose that will work; indeed, a simple 3D model of the world that cannot also take into account additional information such as object weights will not be able to automatically predict such goal poses accurately. Consequently, we require that a "sampling" SMP from ground geometric predicates to their corresponding goal poses be provided by the user. For example, predicate $canMovGr(k1, gr1, curr, pc)$, which checks whether Kuka arm $k1$ combined with gripper $gr1$ can move to a pose from where pallet carrier pc can be grasped, will map to the set consisting of just the single pose depicted in Fig. 1.

We describe SMP as follows. Let $P = \{p_1(o_1, \ldots, o_j), \ldots, p_n(o'_1, \ldots, o'_k)\}$ be the set of ground instances of all geometric predicates occurring in the agent, and $P_s = \{p_1, \ldots, p_n\}$ and $O = \{o_1, \ldots, o_j, \ldots, o'_1, \ldots, o'_k\}$ their associated predicate and constant symbols, respectively. Then, if n_{max} is the maximum parity of a predicate in P, function SMP is denoted by the partial function $\text{SMP} : C \times P_s \times O_1 \times \ldots \times O_{n_{max}} \to 2^C$, where C is the configuration space and each $O_i = O$. Thus, function SMP is a user-defined "sampling" with only the goal poses that "matter" with respect to the current pose $c \in C$ and the given ground geometric predicate. In practice, the full goal pose for a task such as picking up an object could be computed dynamically from a user-supplied pose for the gripper—such as the one in Fig. 1—by first transforming the gripper's pose to "place" it relative to the object and within the current world W, and then using inverse kinematics to derive suitable poses for the geometric bodies that form the robot arm, which are attached to the gripper and to each other.

Function SMP is used within an "intermediate layer" like the ones used in [2, 3], which we actualise here via a special evaluable predicate denoted by

INT : $P_s \times O_1 \times \ldots \times O_{n_{\max}} \to \{true, false\}$, where P_s, $n_{\max}$ and each O_i are as
before. For example, if $n_{\max} = 4$ in the given domain, the agent developer might
invoke the intermediate layer via function $INT(canMov, k1, curr, t1, null)$, where
null is a symbol reserved for unused parameters. Function
$INT(canMov, k1, curr, t1, null)$ is defined as follows. Suppose c_I is the current pose
of the robot, and that SOL ("solution") and FCT ("facts") are global variables
initialised to the empty sequence and empty set, respectively. Then, if there is a
pose $c_G \in SMP(c_I, p, o_1, \ldots, o_{n_{\max}})$, and a collision-free motion plan from c_I to c_G,
we first assign the motion plan to SOL and then return *true*, and otherwise we
assign the set of facts describing why there was no trajectory—specifically the
obstruction(s) that were involved—to FCT and return *false*. This approach keeps
trajectories and poses transparent to the agent developer.

4 Encapsulating Geometric Reasoning Within AgentSpeak

AgentSpeak-like languages offer some useful, built-in mechanisms that allow a
clean embedding of motion planning. In particular, we can encapsulate each geometric predicate $p(v)$ occurring in the agent within a unique achievement goal
$!e_p(v)$ via the plan-rules and action-rules shown below. Specifically, we first
associate the achievement goal with the two plan-rules in the left-hand column
below:

$+!e_p(v) : true \leftarrow actSucc_p(v)$	$actSucc_p(v) : INT(p, v) \leftarrow exec(); post(); \Phi^\top$
$-!e_p(v) : true \leftarrow actFail_p(v)$	$actFail_p(v) : \neg INT(p, v) \leftarrow post(); \Phi^\perp$

Since the bottom plan-rule handles a goal-deletion event, it is only triggered if
the top plan-rule fails, i.e. if the precondition of the ground action $actSucc_p(o)$,
which involves motion planning, is not applicable. Moreover, as per the semantics
of goal-deletion events, once the bottom rule finishes executing, the associated
achievement goal $!e_p(o)$ will still fail. These are the semantics we desire in order to
compute and include (before failing) the beliefs/facts relating to why the failure
occurred. Sets $\Phi^\top$ and $\Phi^\perp$ are predefined beliefs denoting any "predictable"
changes resulting from the achievement goal's execution; e.g., geometric predicate
$canMovGr(K, Gr, curr, PC)$ might have $\Phi^\top = \{r\}$ and $\Phi^\perp = \{\neg r\}$, with $r =$
$reachable(K, Gr, PC)$ (i.e. pallet carrier PC is reachable to arm K with gripper Gr).
The second step in our encapsulation is defined in the right-hand column above
by the action-rules associated with actions $actSucc_p$ and $actFail_p$.[3] In our definition,

---

[3]For simplicity we omit the last parameters of $INT(p, v)$, which may be *null* constants.

post() is a function that returns the set of (symbolic) facts representing either the pose that resulted from executing *exec*(), or the "reasons" why there was no trajectory while evaluating the precondition, i.e. the set FCT computed by $INT(p, o)$. Likewise, *exec*() is associated with a procedure that executes (in the real world) a given motion plan, which in our case is the one that was assigned to SOL when $INT(p, o)$ was called. Action $actFail_p(o)$ is not associated with any such function because its action-rule is only chosen when there is no viable motion plan. Thus, the rule's precondition confirms that $\neg INT(p, o)$ still holds, just in case there was a relevant change in the environment after $INT(p, o)$ was last checked, causing $INT(p, o)$ to now hold (in which case there are no failure-related facts to include).

We assume that *exec*() always succeeds and that if necessary the programmer will check whether the action was actually successful by explicitly testing its desired goal condition. This is exemplified by the $!move(R, F, T)$ achievement goal in Sect. 2, where $?at(R, T)$ checks whether the $navigate(R, F, T)$ action was successful. One property of the described encapsulation is that looking for motion plans and then executing them and/or applying the associated symbolic facts are one atomic operation—no other step can be interleaved to occur between those steps. This ensures that a motion plan found while evaluating an action's precondition cannot be invalidated by an interleaved step while the action is being executed.

Once all geometric predicates have been encapsulated as described, we may then use their corresponding achievement goals from within plan-rules. Since we cannot include them in context conditions (logical formulae) they can instead be placed as the first steps of plan bodies. This allows such achievement goals to be ordered so that the ones having the most computationally expensive geometric predicates are checked only if the less expensive ones were already checked and they were met.

5 Symbolic Abstractions of Geometric Elements

There are certain elements in the geometric representation that are worth abstracting out into their corresponding symbolic entities so that they may be exploited by the agent. Our first abstraction is a user-defined surjection from a subset of the geometric bodies (defined as a sequence of boundary points, for example) onto a subset of the constant symbols occurring in the agent. This allows multiple bodies—such as the individual pieces of a Kuka arm—to simply be identified by a single constant symbol such as $k1$, and also for certain geometric bodies (e.g. an unknown box on the floor) and symbolic constants (e.g. the name of a customer) to be ignored. Indeed, while every rigid body is crucial for geometric reasoning, it does not necessarily need a corresponding symbolic representation, and likewise, every constant symbol occurring in the agent does not necessarily represent a geometric body.

Our second abstraction is represented by logical literals, whose ground instances are obtained and applied via the function $post()$. Formally, these literals are a consistent subset of $2^{P \cup \bar{P}}$, where $\bar{P} = \{\neg p | p \in P\}$ and P is the set of ground instances of predicates occurring in the agent, obtained by replacing each predicate's vector of n terms with an arbitrary vector of n constant symbols. Thus, while these literals will only mention predicate symbols that occur in the agent, they might mention constant symbols (objects) that do not occur in the agent. This leaves room for discovering new, previously unknown objects "on the fly". For instance, if the agent senses from one of its RFID readers that there is a new object on the tool rack, the agent might then look up the tag's associated globally unique electronic product code (EPC) on the web, recognise the object as a certain type of gripper, and assign it with the new symbol $gr7$. This might then become associated with new symbolic facts returned by $post()$, such as $gripper(gr7)$ and $near(gr7, gr1)$.

One useful domain-independent predicate inferable from the geometric representation concerns pairs of bodies that are "attached" to one another in the geometric model. For example, suppose that the vision test in the testing station builds a detailed 3D model of the partially assembled hinge on the pallet carrier, and then checks that it was assembled correctly. If this test fails because a part (e.g. one of the leaves) is absent in the partial hinge, facts such as $att(pc, leaf1)$ and $att(leaf1, retainer)$, indicating which pairs of parts are nonetheless successfully attached to each other in the partial hinge, will enable the agent to reason about which parts will move together when the pallet carrier is transferred onto the conveyer belt. Formally, a possible definition of $att(o, o\prime)$ for two objects o, o' is the following (we use A, A' and $C^A, C^{A'}$ in C to respectively denote their bodies and configuration spaces): $att(o_1, o_2)$ holds if there is a $k \in \mathbb{R}$ and $m \in \{1, \ldots, 3\}$ such that for any two poses $(a_1, a_2, a_3, a_4) \in C^A$ and $(a_1', a_2', a_3', a_4') \in C^{A'}$, we have $a_m = a_m' + k$, i.e. at least one degree of freedom of one of the objects is constrained by the other. Other useful domain-independent predicates include $vol(o, v)$ and $coll(o, o')$, where the former is the volume v of object o calculated from its geometric representation, and the latter indicates that there is a pose in which o and o' (e.g. the two arms) will collide; formally, $coll(o, o')$ holds if there exist bodies A, A' associated respectively with objects o, o', and poses $c \in C^A$ and $c' \in C^{A'}$ such that $A(c) \cap A'(c') \neq \emptyset$, i.e. when A, A' are transformed and 'placed' into world W, at least one of their points overlap. Such a fact might eventuate in the agent taking precautions to ensure the tool rack is only used by one arm at a time.

There are also geometric elements that are too 'fine grained' to be modelled as symbolic elements, such as absolute x and y coordinates, and orientations of objects in 3D space; doing so may well lead to an explosion in the symbolic state space [3]. Moreover, as pointed out in [2], there are also relevant symbolic facts that do not depend on a pose, such as the number of products assembled so far and the weight of a new part. These facts can be managed directly by the agent, for example by directly sensing the environment.

In the situation where there was no viable motion plan when the precondition of an action-rule above was checked, the facts applied by $post()$ instead "describe" the reason. To this end, two useful domain-independent predicates, inspired by [3], are $obsSome(k2, canMov, k1, t1)$, indicating arm $k2$ obstructs at least one trajectory of the task $canMov(k1, t1)$, and likewise $obsAll(k2, canMov, k1, t1)$. The agent could exploit such information by, for instance, moving arm $k2$ out of the way.

6 Conclusions and Future Work

We have presented an approach to interfacing BDI agent reasoning with geometric planning in a way that uses some of AgentSpeak's existing constructs and semantics. We have also shown how interesting abstractions can be extracted from the detailed geometric model and then exploited during agent reasoning. We intend to further study these abstractions, e.g. how to compute $obsSome(k2, canMov, k1, t1)$, and to formalise the integration by extending the operational semantics of AgentSpeak.

Acknowledgements We thank Amit Kumar Pandey and the reviewers for useful feedback. Felipe thanks CNPq for support within grant no. 306864/2013-4 under the PQ fellowship and 482156/2013-9 under the Universal project programs. The other authors are grateful for support from the Evolvable Assembly Systems EPSRC project (EP/K018205/1), and the PRIME EU FP7 project (Grant Agreement: 314762).

References

1. Rao, A.S.: AgentSpeak(L): BDI agents speak out in a logical computable language. In: Proceedings of the MAAMAW Workshop, pp. 42–55 (1996)
2. de Silva, L., Pandey, A.K. Alami, R.: An interface for interleaved symbolic-geometric planning and backtracking. In: IROS, pp. 232–239 (2013)
3. Srivastava, S., Fang, E., Riano, L., Chitnis, R., Russell, S., Abbeel, P.: Combined task and motion planning through an extensible planner-independent interface layer, ICRA, pp. 639–646 (2014)
4. Lagriffoul, F., Dimitrov, D., Saffiotti, A., Karlsson, L.: Constraint propagation on interval bounds for dealing with geometric backtracking, IROS, pp. 957–964 (2012)
5. Erdem, E., Haspalamutgil, K., Palaz, C., Patoglu, V., Uras, T.: Combining high-level causal reasoning with low-level geometric reasoning and motion planning for robotic manipulation, ICRA, pp. 4575–4581, (2011)
6. Plaku, E. Hager, G.D.: Sampling-based motion and symbolic action planning with geometric and differential constraints, ICRA, pp. 5002–5008 (2010)
7. Dornhege, C., Eyerich, P., Keller, T., Trüg, S., Brenner, M. Nebel, B.: Semantic attachments for domain-independent planning systems, ICAPS, pp. 114–121 (2009)
8. Gaschler, A., Kessler, I., Petrick, R., Knoll, A.: Extending the knowledge of volumes approach to robot task planning with efficient geometric predicates. ICRA, (2015) (To appear)

9. Kaelbling, L.P., Lozano-Pérez, T.: Integrated task and motion planning in belief space. IJRR 32(9–10), 1194–1227 (2013)
10. LaValle, S.M.: Planning Algorithms. Cambridge University Press (2006)
11. Antzoulatos, N., Castro, E., de Silva, L., Ratchev, S.: Interfacing agents with an industrial assembly system for "plug and produce". In AAMAS, pp. 1957–1958 (2015)

A Holonic Manufacturing System for a Copper Smelting Process

Carlos Herrera, José Rosales, André Thomas and Victor Parada

Abstract In this article, a holonic manufacturing system to coordinate the activities in a copper concentrate smelting process is designed. The system was developed and compared with a simulation of the current system in use. Four scenarios with different feeding levels of copper concentrate were considered. The results show an increase in the amount of processed ore in the four studied scenarios because of a reduction in the total waiting time and more efficient use of the plant equipment. Thus, the better use of resources and a decrease in idle time because of improved coordination of activities and different components of the system is highlighted.

Keywords Manufacturing · Copper smelter · Holonic manufacturing systems · Multiagent simulation · Metallurgical processes

C. Herrera (✉) · J. Rosales
Departamento de Ingeniería Industrial, Universidad de Concepción,
Concepción, Chile
e-mail: cherreral@udec.cl

J. Rosales
e-mail: jrosalesl@udec.cl

A. Thomas
Centre de Recherche en Automatique de Nancy, Université de Lorraine,
Nancy, France
e-mail: andre.thomas@univ-lorraine.fr

V. Parada
Departamento de Ingeniería Informática, Universidad de Santiago de Chile,
Santiago, Chile
e-mail: victor.parada@usach.cl

© Springer International Publishing Switzerland 2016   189
T. Borangiu et al. (eds.), *Service Orientation in Holonic and Multi-Agent Manufacturing*, Studies in Computational Intelligence 640,
DOI 10.1007/978-3-319-30337-6_18

1   Introduction

Production control is a critical aspect in metallurgical production systems integrating continuous and discrete processes. A production system with these characteristics, which are fundamental for copper production, corresponds to the smelting process. This process involves three fundamental stages: smelting, conversion and refining of the ore. The copper concentrate is continuously loaded into the smelting furnace, whereas the material loading is discrete in the following stages. Specifically, the ore must be transported from one stage to another using ladles, which are led by a bridge crane. In practice, the three metallurgical processes attract most attention, whereas the material handling of ores between stages is not prioritized [1, 2].

Because of its focus on the physical processes of production systems, a Holonic Manufacturing System (HMS) is presented as a suitable alternative to address the coordination problem between continuous and discrete processes [3–6]. This kind of systems correspond to the paradigm of designing production systems based on a distributed organization of autonomous intelligent and collaborative entities called holons. The holons work together to increase the flexibility of the system. In practical terms, such systems are implemented as software modules that interact on a computational platform, and each module represents physical entities/activities such as machining, processes, material handling devices, etc. Thus, a HMS provides a natural and efficient method to manage dynamic and complex systems. Furthermore, these systems have been widely proposed as a solution for flexibility, reactivity, modularity, scalability and autonomy in manufacturing systems [4].

The problem of metallurgical control has been typically addressed using expert systems and decision support systems [7, 8]. These systems model the production planning by considering the decision-making process in a centralized manner. Systems with continuous and discrete variables appear too complex and too time consuming to be simulated under these techniques. In these cases, it appears more appropriate to tackle the system dynamics with distributed decisions specialized for the system to manage the continuous and discrete aspects of the process. With this approach, it is possible to obtain greater flexibility and reactivity, and to achieve an effective configuration [9].

Several studies have shown the advantages of using HMS in chemical processes. The application of a HMS in the chemical processing industry has given rise to the "Holonic Process Plant" concept [10]. The results show that a holonic approach will address problems such as the lack of flexibility and reconfigurability of this kind of industrial system. Also in this field, the holonic control of a water-cooling system for a steel rod mill has been proposed as part of a feasibility study for the initiative of Intelligent Manufacturing Systems (IMS) [11]. The proposed architecture was simulated to prove the benefits of a holonic system versus a conventional control system. Both works focus on the application of IMS to continuous processes. Recently, agent-based service-oriented integration architecture for automating the chemical process has been proposed to quickly respond to changes and failures while considering on going processes [12]. Unlike the chemical industry, metal-lurgical processes are more structurally complex because discrete and continuous events occur simultaneously.

In this paper we propose a decentralized control architecture. The design of a holonic system to manage the production of a copper smelter is presented. The set of designed modules was computationally implemented using the SPADE (Smart Python multi-Agent Development Environment) tool [13] and Discrete Events Simulation (DES) software Rockwell Arena® 14.7 [14]. The resulting platform allows simulating the operations of the holonic system and study the coordination among the activities in the copper smelter. Thus, it is possible to decrease the waiting times and better distribute the workload among different stages of the process.

2 Proposed Holonic System

2.1 Copper Smelting Process

Copper is purified on a large scale to obtain metallic copper using a metallurgical process in a copper smelter. The extracted ore from a mine is transported to a plant, where it is ground and partially purified to reach a purity of approximately 30 %, which is called copper concentrate. Three processes are considered in the metal-lurgical process. The first stage corresponds to the mineral smelting in furnaces at high temperatures, at which the mineral is found in the liquid phase. As a result of the chemical reaction at that temperature, a mixture of 60–75 % of purity is gen-erated. Then, a bridge crane transports the ladles with the mineral to the second conversion stage. In this stage, a set of chemical reactions purifies the copper to a purity level of approximately 99.0 %. In the final refining stage, the remnants of sulphur, arsenic and oxygen are removed, and a purity of 99.5 % is obtained. Subsequently, the product is poured into casting wheels, cooled and shaped into copper plates.

The production system that we studied consists of a smelting furnace, four converters and two bridge cranes (Fig. 1). The copper concentrate is continuously fed to the smelting furnace, and the obtained product is loaded into ladles and

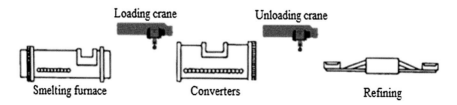

Fig. 1 Copper smelting process

transported by the crane to the converters. The conversion process is performed in four converters, which operate in batch mode, i.e., the converters are loaded at the beginning of the cycle and the conversion is performed during a cycle of several hours. Then, the resulting product is transported in ladles by a second crane. Because of the constraints of the metallurgical process, only two converters operate simultaneously. A third converter remains ready to operate, and the fourth converter is under maintenance. This study considers material handling from the smelting furnace to the converters.

2.2 Holonic-System Design

The proposed HMS is based on the communication and coordination of several holons which correspond to the resources of the system: smelting furnace, converters and cranes. The holons contain the logic to perform their operations and reach their local goals. A diagram of the activities of the HMS logic is presented in Fig. 2. The following holons and agents were designed:

- **Converter holon**: each holon that represents a converter contains information about its capacity, its status, and the status of the material loading and unloading line. Each converter receives information on the status of the cranes and the amount of material contained in the smelting furnace. Also, the holons can calculate the estimated ending time of an activity. The converters report to the cranes their status, maximum loading level and current level.
- **Crane holons**: represent the cranes of the system and contain information about their status and the converter to which they are assigned during an operation. The cranes receive information on the status of the converters and the amount of material contained in the smelting furnace. They report their status to converters.
- **Furnace holon**: this holon sends the information about the current amount of available material to be processed. It can regulate its production within maximum and minimum limits.
- **Messenger agent**: this is an intermediate agent between the simulation and the multi-agent system that receives information from the simulation software and sends it to the corresponding holon. Similarly, the information received from each holon is sent to the simulation software. Because of the software limitations, the sent messages are not too close to one another. In practice, it has been verified that the acceptable time between messages is one second.

2.3 HMS Simulation

The simulation model consists of three modules that comprise the logic of the system: the converter, production and crane modules. The first module generates

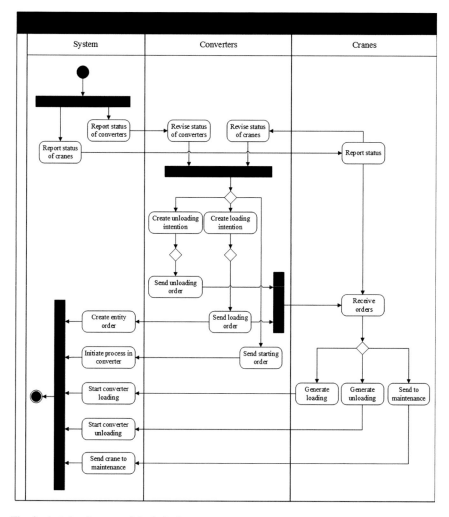

Fig. 2 Activity diagram of the holonic system

the production order for the system. The first available converter is selected to load the material, and the constraint that no more than two converters are simultaneously in operation is verified. After unloading is finished, the converter is available for the entry of a new order. To start the simulation, two entities that represent two cranes in the system are generated. These cranes wait for a loading or unloading order from the converter module. After the order is received, the entity is input to the corresponding process. The loading status of the converter where the process is performed is revised, and it finishes when the filling or emptying objective of the converter is attained. Then, the crane is sent to perform an external process if no crane has already performed it. Finally, the crane returns to wait for a new loading or unloading request.

The computational implementation of the HMS was performed using the SPADE platform. This software is free under the GNU General Public License. To simulate the smelting process and send data on the holon status to the HMS, the software Arena® Simulation 14.7 was used. A library in the SIMAN C++ language provided by Arena® was used to implement the inter-process communication.

The operation of the copper smelter was simulated for two cases: current operational activities (existing control) and operations controlled by the proposed system (holonic control). For both cases, the simulation horizon was one week. In total, 13 replicas of the existing control and two of the holonic control were performed.

For each case, four different scenarios were simulated with various feedings of copper concentrate into the smelting furnace: 67.6, 73.8, 80.3 and 86.5 ton/h. These levels have been defined by the company according to its upstream constraints and operational rules.

3 Results

The control of the copper smelter by the holonic system improves the amount of processed copper concentrate in the four studied scenarios. This improvement is observed at both daily (Q_{ccd}) and weekly (Q_{ccw}) levels (Table 1). It should be noted that the copper concentrate was similarly processed in the four simulated scenarios. In particular, scenarios 3 and 4 had identical values. Identical phenomena are observed when the control is performed with the holonic system. In the latter case, scenarios 2 and 4 are equal. In addition, the amount of copper concentrate processed in one week did not vary in the four scenarios. On average, the amount of processed concentrate copper was increased between 7.4 and 9.0 % in the four scenarios.

These results suggest that the control by the holonic system enables an efficient management of the production process and increases the copper concentrate processing capacity. Then, the implemented rules contribute to streamlining operations and decreasing the critical time of the production process.

Comparing the holonic control with the existing control, the amount of processed copper concentrate is increased because of a reduction in total waiting time

Table 1 Amount of processed copper concentrate and waiting times

	Scenario 1 67.6 ton/h		Scenario 2 73.8 ton/h		Scenario 3 80.3 ton/h		Scenario 4 86.5 ton/h	
	Existing sim.	Holonic sim.	Existing sim.	Holonic sim.	Existing sim.	Holonic sim.	Existing sim.	Holonic sim.
Q_{ccw} (ton)	9849.6	10,608.0	9816.0	10,576.8	9878.4	10,764.0	9878.4	10,608.0
Q_{ccd} (ton)	1407.1	1515.4	1402.3	1511.0	1411.2	1537.7	1411.2	1515.4
t_{wp} (min)	197.1	67.5	195.6	70.2	196.9	72.1	198.2	64.1
t_{wu} (min)	82.3	53.2	82.3	56.6	82.4	54.1	82.4	56.7

when the copper smelter is controlled by the holonic system. In Table 1, the waiting time to process (t_{wp}) and waiting time to unload from the converters (t_{wu}) are presented. In the four scenarios, it is verified that both waiting times decrease when the holonic control is simulated; t_{wp} and t_{wu} decrease by 64.9 and 32.8 %, respectively, because in the existing control, the level significantly decreases with the loading of the first converters and subsequently maintains a cyclic behaviour in time. On the contrary, in the holonic control, more time is required to reach the cyclic behaviour. Specifically, the furnace reaches its maximum capacity after 40 h, which is double the required time for the existing control.

The production level is also improved because the copper smelter is more efficiently used. The coordination of the assigned tasks to the cranes is important because most of the delays in the current operational activities arise from the waiting time of both loading and unloading of the material from the converters or the furnace. The use of cranes corresponds to the percentage of time in relation to the total simulation time, in which a crane performs some process (loading, unloading or external process). In Table 1, the percentage differences in the use of cranes between existing and holonic controls are given. The increase in use is higher for crane 2 than crane 1 because the initial use of crane 2 facilitates its assignment to specific external tasks of the process, such as the transport of slags which are removed during the conversion process. Table 2 shows that the first converter was used 15.93 % more than in the holonic control. Converters 1 and 3 were more used, whereas converter 2 was less used. These results show that the utilization of the converters was homogenized by reducing idle times.

The overall performance of both controls is shown in Fig. 3; the use of three converters for the existing control is visualized, while the holonic control is observed in Fig. 3b. In the first case, converter 1 (of greatest capacity) is loaded. After this stage is completed, the loading of the second converter proceeds. Immediately after both converters finish loading, processing begins. Converter 1 is continuously active, whereas there are intermittencies in the other two converters. Because only two converters can operate simultaneously, the first process of the third converter only begins after converter 1 completes its process stage. Figure 3b suggests that in the holonic control the loading process of the converters is more homogenously performed reducing idle times and more quickly finishes the processes.

Table 2 Difference in use between existing and holonic control (%)

Equipment	Scenario 1	Scenario 2	Scenario 3	Scenario 4
Crane 1	9.85	7.64	8.19	8.08
Crane 2	24.24	23.56	24.25	23.83
Conv. 1	15.93	12.25	13.62	14.15
Conv. 2	−11.49	−7.49	−6.06	−11.92
Conv. 3	12.93	16.97	19.40	14.01

Fig. 3 Processing cycles in
the converters

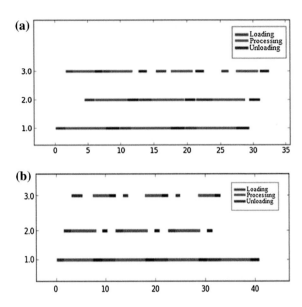

4   Conclusion

In this article a holonic system to control and manage material in a copper smelter is
presented. We analysed the material handling from the smelting furnace to the
converters and subsequently to the refining stage. The production system was
simulated to evaluate the proposed holonic system versus the existing one.

The implementation of the holonic system decreases the waiting times and better
distributes the workload among different machines. This improvement was
achieved by coordinating the components, making decisions in real time and
adapting to any inconvenient or difference in the optimal process without the
intervention of a third party. In addition, this distributed system was verified to
present sufficient flexibility adapted to this complex production process.

Acknowledgments The authors would like to thank the Complex Engineering Systems Institute,
ICM: P-05-004-F, CONICYT: FBO16, DICYT: 61219-USACH, ECOS/CONICYT: C13E04,
STICAMSUD: 13STIC-05.

References

1. Pradenas, L., Zúñiga, J., Parada, V.: CODELCO, Chile programs its copper-smelting
   operations. Interfaces **36**(4), 296–301 (2006)
2. Pradenas, L., Campos, A., Saldaña, J., Parada, V.: Scheduling copper refining and casting
   operations by means of heuristics for the flexible flow shop problem. Pesqui. Oper. **31**(3),
   443–457 (2011)

3. Blanc, P., Demongodin, I., Castagna, P.: A holonic approach for manufacturing execution system design: an industrial application. Eng. Appl. Artif. Intell. **21**(3), 315–330 (2008)
4. Babiceanu, R.F., Chen, F.F.: Development and applications of holonic manufacturing systems: a survey. J. Intell. Manuf. **17**(1), 111–131 (2006)
5. Arauzo, J.A., Del-Olmo-Martinez, R., Lavios, J.J., De-Benito-Martin, J.J.: Scheduling and control of flexible manufacturing systems: a holonic approach. Rev. Iberoamer. Autom. E Inform. Ind. **12**(1), 58–68 (2015)
6. Valckenaers, P., van Brussel, H., Bongaerts, L., Wyns, J.: Holonic manufacturing systems. Integr. Comput. Aided Eng. **4**(3), 191–201 (1997)
7. Carayannis, G.: Artificial intelligence and expert systems in the steel industry. JOM **45**(10), 43–51 (1993)
8. Zhao, Q.-J., Cao, P., Tu, D.-W.: Toward intelligent manufacturing: label characters marking and recognition method for steel products with machine vision. Adv. Manuf. **2**(1), 3–12 (2014)
9. Herrera, C., Belmokhtar-Berraf, S., Thomas, A., Parada, V.: A reactive decision-making approach to reduce instability in a master production schedule. Int. J. Prod. Res. 1–11 (2015)
10. Chokshi, N.N., McFarlane, D.C.: Rationales for holonic manufacturing systems in chemical process industries. In: 2012 23rd international workshop on database and expert systems applications, vol. 1, p. 616, Los Alamitos, CA, USA (2001)
11. Agre, J.R., Elsley, G., McFarlane, D., Cheng, J., Gunn, B.: Holonic control of a water cooling system for a steel rod mill. In: Proceedings of the Fourth International Conference on Computer Integrated Manufacturing and Automation Technology, pp. 134–141 (1994)
12. Luo, N., Zhong, W., Wan, F., Ye, Z., Qian, F.: An agent-based service-oriented integration architecture for chemical process automation. Chin. J. Chem. Eng. **23**(1), 173–180 (2015)
13. https://pypi.python.org/pypi/SPADE
14. https://www.arenasimulation.com/

A Nervousness Regulator Framework for Dynamic Hybrid Control Architectures

Jose-Fernando Jimenez, Abdelghani Bekrar, Damien Trentesaux
and Paulo Leitão

Abstract Dynamic hybrid control architectures are a powerful paradigm that addresses the challenges of achieving both performance optimality and operations reactivity in discrete systems. This approach presents a dynamic mechanism that changes the control solution subject to continuous environment changes. However, these changes might cause nervousness behaviour and the system might fail to reach a stabilized-state. This paper proposes a framework of a nervousness regulator that handles the nervousness behaviour based on the defined nervousness-state. An example of this regulator mechanism is applied to an emulation of a flexible manufacturing system located at the University of Valenciennes. The results show the need for a nervousness mechanism in dynamic hybrid control architectures and explore the idea of setting the regulator mechanism according to the nervousness behaviour state.

Keywords Nervousness · Dynamics · Hybrid control architecture · Switching · Multi-agent system

J.-F. Jimenez (✉) · A. Bekrar · D. Trentesaux
LAMIH, UMR CNRS 8201, University of Valenciennes and Hainaut Cambrésis, UVHC,
59313 Le Mont Houy, France
e-mail: j-jimenez@javeriana.edu.co

A. Bekrar
e-mail: abdelghani.bekrar@univ-valenciennes.fr

D. Trentesaux
e-mail: damien.trentesaux@univ-valenciennes.fr

J.-F. Jimenez
Pontificia Universidad Javeriana, Bogotá, Colombia

P. Leitão
Polytechnic Institute of Bragança, Bragança, Portugal
e-mail: p.leitao@ipb.pt

P. Leitão
LIACC—Artificial Intelligence and Computer Science Laboratory, Porto, Portugal

© Springer International Publishing Switzerland 2016   199
T. Borangiu et al. (eds.), *Service Orientation in Holonic and Multi-Agent
Manufacturing*, Studies in Computational Intelligence 640,
DOI 10.1007/978-3-319-30337-6_19

1 Introduction

A dynamic hybrid control architecture (D-HCA) is a promising control model that adapts to complex system demands. ADACOR [7] and PROSA [16] are two of the most known approaches in D-HCA. In general, these architectures feature a reconfiguration property that adjusts the functioning of the system by tailoring the control solution according to the corresponding system needs. However, the dynamic characteristic of these architectures challenges the efficiency of this paradigm. During the reconfiguration process, the system may experience some instability resulting from an improperly synchronized evolution process [8]. In particular, this process lacks sufficient time to stabilize the solution and activate the benefits from the new configuration [15]. In this situation, the system is undergoing nervousness behaviour. For this paper, *nervousness behaviour* is a conduct of a whole or part of a system in which its decisions or intentions change erratically without leaving a sufficient time for stabilizing into an expected functioning. For example, in the flexible manufacturing system, the degree of nervousness behaviour in the system increases when the products change constantly its intentions of reaching a machine in an assembly system. Thus, it is crucial to control the nervousness behaviour of a system to avoid the unstable and chaotic behaviour in D-HCAs.

The paper proposes a framework of a nervousness regulator that handles the nervousness behaviour based on an indicator of nervousness. In addition, a classification of nervousness is created that responds to the necessity of dealing with different mechanisms according to the current nervousness behaviour of the system. The paper is organized as follows: Sect. 2 reviews the nervousness behaviour in dynamic systems. Then, Sect. 3 presents the general framework of a nervousness regulator for D-HCA. An instantiation of the framework applied to a case study of a flexible manufacturing system is presented in Sect. 4. Section 5 describes an experimental case study and illustrates the need for a nervousness regulator in D-HCA. Finally, Sect. 6 concludes the findings of this research and provides future research to be addressed.

2 Nervousness Behaviour: Literature Review

Nervousness behaviour in dynamic systems has several definitions. Initially, a nervous system was used to describe the perturbations occurring in the material requirement planning (MRP) systems. In this domain, the concept of a nervousness behaviour started as the changes of intentions, experienced by low-level entities, when the master schedule does not change significantly [13]. In this case, the nervousness is defined by the difference between the planning and the real execution of operations. Subsequently, the term evolved to the instability derived from internal/external changes or perturbations that cause the system process to be treated as an exception [2, 9]. However, the changes occurring on internal and external levels can be either great enhancement opportunities or extremely disruptive events for the system's behaviour [11].

In the D-HCA domain, the nervousness behaviour is present in the changes made in the control solution. In this sense, a D-HCA that avoids the nervousness behaviour must balance the tension of performing sufficient changes for reacting and enhancing the system's performance while maintaining a stable and safe evolution [1, 10]. In general, the condition that a system experiences nervousness behaviour is not a negative comportment. However, it is crucial to dampen the nervousness in order to avoid experiencing a nervousness state. When the system presents nervousness behaviour, the system reacts to internal and external stimuli erratically [3] and increases a non-coordinated solution towards the achievement of system objectives.

Thus, researchers have introduced the nervousness regulator to dampen this behaviour. Hadeli et al. [3] proposed a mechanism that assures that the perceived improvement in the system's evolution is good enough. The dampening of the nervousness is preventive and affects directly to change of agent's intentions. Another way to detect nervousness is to evaluate the tendency of a specific performance indicator [4, 5]. The authors propose a mechanism that monitors a nervousness indicator in different time windows. In this approach, the nervousness behaviour is avoided by monitoring the tendency of an indicator. As another example, Barbosa et al. [1] propose a nervousness control mechanism based on a proportional, integral and derivative (PDI) feedback controller to support the system dynamism. In this case, the threshold is not fixed and, due to the degree of nervousness, the regulator intervenes to keep the system's stability.

It can be seen from the literature reviewed that the nervousness behaviour is concentrated in the changes of agents in distributed architectures. Certainly, these changes influence the emergence behaviour of the system. However, the changes in the control architecture of the system's structure and its behaviour have not been properly explored simultaneously. Additionally, although researchers have focused on the detection and handling of the nervousness behaviour, very few solutions have been proposed to prevent and mitigate the occurrence of such behaviour. Therefore, we propose to include a nervousness regulator in a D-HCA that damps the instability derived from continuous structure/behaviour changes; a framework is conceived that partitions the regulation strategy according to the degree of nervousness.

3 Proposed Nervousness Regulator Framework in D-HCA

This section presents the proposed framework to control the nervousness behaviour in a D-HCA. The framework identifies four phases of the nervousness behaviour according to the nervousness state: prevention, assessment, handling and recovery. Figure 1 illustrates the framework's phases.

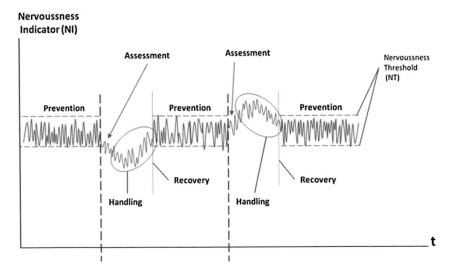

Fig. 1 Phases of a system towards nervousness behaviour

The management of the nervousness behaviour is based on monitoring of the dynamic systems with a **nervousness indicator** (**NI**). This indicator shows the nervousness of the system by measuring the system stability. Additionally to this indicator, this framework defines a **nervousness threshold** (**NT**) as the maximum nervousness level allowed by the system during execution. Once the threshold is passed, the nervousness regulator starts functioning. This out-of-control event is addressed differently depending on the state of the nervousness behaviour. A framework consists of an iterative four-phase method used to control the nervousness behaviour of a system. Each phase is explained in Table 1.

Table 1 Phases of the system towards a nervousness behaviour

Nervousness crisis prevention:	Nervousness crisis assessment:
1. Mitigate the risk of nervousness behaviour.	1. Discover the nervousness incident.
2. Establish auto-regulators that monitors and autocorrect the nervousness indicator.	2. Establish evaluation methods that detect the incident by the infringement of threshold.
3. Conditional-rules or heuristics.	3. Statistic, data mining or forecasting methods.
Nervousness crisis recovery:	Nervousness crisis handling:
1. Stabilize the system and provide feedback to the prevention phase.	1. Handle and calm the nervousness incident by reconfiguring.
2. Establish a method to return to normal conditions	2. Establish methods to damp the incident.
3. Feedback and re-tuning parameters (threshold)	3. Simulation-optimization and tuning methods

In D-HCA, the nervousness present in the system is due to the changes performed by a switching mechanism that modifies or reconfigures dynamically the structure and/or behaviour of the system to obtain a custom-built optimal configuration. However, in order to accomplish this objective, the system might switch constantly causing a nervousness event. In this paper, we focus on the nervousness crisis prevention phase of the nervousness regulator in order to prove the need for a mechanism in the system. In this respect, the nervousness control authorises or not the switching procedure depending on the nervousness threshold. An instantiation of the proposed framework focused on the prevention of the nervousness state is proposed in the next section.

4 Nervousness Regulator of a Flexible Manufacturing System

In this paper, a D-HCA of a flexible manufacturing system is modelled with a nervousness regulator. At first, the D-HCA constructed specifically for the case study is presented. Then, the inclusion of the nervousness regulator in the defined D-HCA is described. The D-HCA is modelled as a dispatching scheduler with an agent-based solution. While the scheduler is a MILP solution for a flexible job shop problem only for dispatching, the jobs are intelligent entities represented by agents within the simulation. The reason for using this solution responds to the idea of giving full autonomy to the agents for monitoring the changes of intentions during product execution.

4.1 The Case Study of a Flexible Manufacturing System

The manufacturing system in the paper corresponds to a flexible manufacturing system (FMS) located at the University of Valenciennes (France) in the AIP PRIMECA lab. It consists of seven workstations (M1, M2, M3, M4, M5, M6, M7) connected by a flexible transportation system. Seven different assembly jobs (B, E, L, T, A, I and P) can be produced in the FMS and each has a sequence of operations including O1, O2, O3, O4, O5, O6, O7 and O8 to be executed. Each workstation can perform a subset of operations Oi. The production starts when a holding-case is loaded in M1 in the moving shuttles for O7. Once the sequence of each job has been processed, the shuttle returns to M1 to be unloaded in operation O8. The AIP PRIMECA facility is modelled as a flexible job shop problem (FJSP) with material processing and handling flexibility. The manufacturing system layout, the sequence of operations for each job and the processing times for each operation in the AIP PRIMECA are available in the benchmark of Trentesaux et al. [14].

4.2   D-HCA of the Flexible Manufacturing System

The D-HCA of this paper is based on the governance mechanism approach proposed in Jimenez et al. [6]. This approach features an operating mode of a D-HCA as a specific parameterization that characterizes the control settings applied to the system. A switching mechanism, called governance mechanism, commutes the operating mode to reconfigure the architecture of the control system. The D-HCA that controls the FMS is organized as follows (Fig. 2):

FMS Controlled system: the general structure of the FMS is divided into two layers: a global and a local layer. While the global layer contains a unique global decisional entity (GDE) responsible for optimizing the release sequence of the production orders (scheduler), the local layer contains several local decisional entities (LDE) as jobs to be processed in the production order (7 jobs in scenario A0). In this approach, each decisional entity (GDE or LDE) includes its own objective and governance parameters. In this scenario, the objectives of the GDE and LDE are respectively to minimize the makespan at batch execution level and the completion of the next operation. The governance parameter in the GDE is the role of the entity for establishing the order release sequence and imposing these intentions to the LDE in the shop floor. The governance of each LDE is represented by the reactive technique that guides the evolution of the job through the shop floor. This evolution can be driven by a potential-field's (PF) approach [12] or by the first available machine rule (FAM). For this research, even though both PF and FAM techniques are part of the reactive approach in distributed systems, it is considered that the potential-field's approach assures higher performances while computing resource allocation depending on their availability and shortest route to the resources. For a better representation of the configuration, an operating mode vector that gathers all the governance parameters of the decisional entities is defined.

Governance mechanism entity: this switching mechanism is responsible for changing the governance parameters of the GDE and LDE through the operating

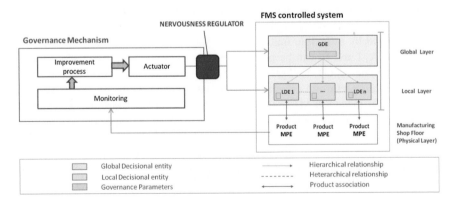

Fig. 2 D-HCA of a FMS with a nervousness regulator

mode vector. It monitors the performance of the controlled FMS, continues with the improvement process for enhancing the system performance and triggers a change in the system's functioning by acting upon the operating mode vector (Fig. 2). Considering that the nervousness behaviour derived from the switching of the control system is monitored, the switching is triggered periodically (every 20 time-units) according to a condition-rule applied to the system. For measuring the performance of operating modes, the expected makespan without switching (*static*) was simulated for each possible operating mode vector. The result was sorted in a numbered list and plotted to characterize the operating modes (Fig. 3 top). The list contains 128 operating modes derived from the combination of the governance parameters of all LDE (*jobs to be produced*). In this model, it is assumed that this characterization of operating mode does not change through the execution and the results are considered a preliminary possible control solution. Finally, the direction of the switching towards an operating mode is decided by a condition-rule according to the intentions received from the resources. That is, if a certain resource has more than α (Alpha) jobs to be produced at the switching time, the operating mode changes to a more reactive one (higher in the numbered list) with a step of λ (Lambda) in the sorted list of operating modes. Otherwise, if all resources have less than four jobs intentions to be produced, the operating mode switches to better alternatives (lower in the numbered list).

Nervousness regulator: This entity is responsible for filtering the intentions of the governance mechanism to dampen the switching evolution. For the definition of nervousness indicator (NI) and Nervousness threshold (NT), the module proposed by Hadeli et al. [3] was used. This module employs a probabilistic mechanism each time the system is willing to change. As it is not evaluating the state of the system but dampening the system evolution, this approach is enclosed in the nervousness crisis prevention in the defined framework. The NI defined is a random value between 0 and 1, and the NT is fixed to β (beta). If NI is higher than NT the system holds the switch. Otherwise, the switching process is performed. The flow diagram of the nervousness linked to the switching mechanism is illustrated in Fig. 3 bottom.

5   Experimental Study and Results

This section presents the experiments performed in the manufacturing cell of the AIP PRIMECA lab. of the UVHC. The main goal of this experiment is to compare the behaviour of a D-HCA with and without a nervousness regulator; we wanted to prove that the nervousness mechanism damps the switching process and avoids thus a nervous behaviour. For the implementation, the proposed D-HCA with nervousness regulator is programmed in the NetLogo agent-based software [17]. The data-set used for the case study is the scenario A0 from the Benchmark [14].

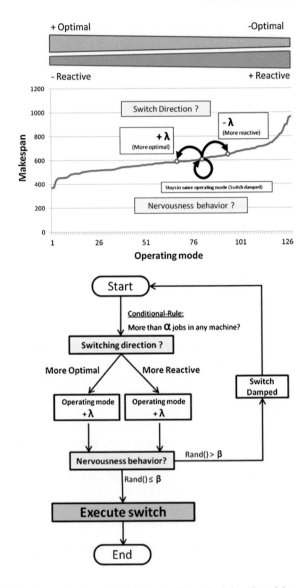

Fig. 3 D-HCA of a FMS with a nervousness regulator

For the setup of the D-HCA, the governance parameters of the decisional entities are initially fixed. The GDE presents a coercive role and the LDE is fixed with the values of the 80th operating mode. As initial values for the experiments, the conditional-rule α is 4, the switching step λ is 2 and the nervousness threshold β is 0.9. When execution starts, the GDE communicates a coercive optimal plan to the LDE for the order release sequence. The emulation of the production system starts execution with this optimal plan and the initial operating mode. In the experiments, while part A considers the proposed D-HCA without the nervousness regulator, part B includes the regulator. Considering that the nervousness regulator is a probabilistic mechanism, it is executed 30 times for each part of the experiment. Finally, an analysis of variance (ANOVA) procedure is conducted to compare the differences between the results of part A and part B.

As a first result, the experiments showed that there are statistically significant differences between part A and B as determined by one-way ANOVA (F $(1,58) = 4.0068, p = 0.05$). In this respect, part B performs better in the production execution. In fact, even though this result does not demonstrate that constant switching can generate a nervousness state, the results show that the nervousness mechanism damps the switching. We believe that the results are essentially by two reasons. The first reason is that, due the rapid evolution of the system, damping the switching is imposing the system to stay in the same operating mode to take advantage of the benefits inherent to the configuration. Thus, the jobs are able to apply certain intentions settled by the operating mode of current execution. The second reason is that, when the nervousness regulator is activated, the jobs enter a stabilization period in which the regulator contributes avoiding the changes of intentions caused by the switching. Even though it was not confirmed in this experiment, the changes of intentions should diminish as a consequence of the stabilization period. These experiments confirm that the switching between different operating modes in a D-HCA achieves a better performance than a fully static configuration. In Fig. 4, while in a static operating the proposed architecture has 607 time-units as makespan, the switching for part A and B presents a mean of 509.40 and 473.76, respectively. In conclusion, from the experiments conducted, the nervousness regulator searches a convergence in the dynamic process in order to stabilize the trade-off between evolution and nervousness. However, these results raise the further need to balance between the switching and the nervousness behaviour mechanisms.

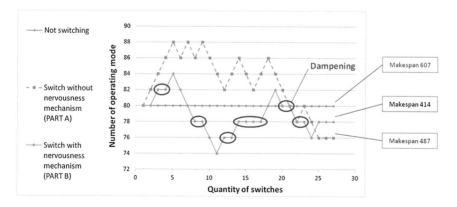

Fig. 4 Examples of the evolution of switching during execution

6 Conclusions

A framework for a nervousness regulator in D-HCA was proposed. The framework defines the prevention, assessment, handling and recovery phases as possible locations to drive the nervousness behaviour present in a dynamic system. An instantiation of the nervousness regulator included in the D-HCA of a manufacturing was tested in an experiment of assembly. Results show that the nervousness regulator is needed to defuse the consequences of nervousness behaviour. The research perspective derived from this paper is to explore different models handling the nervousness and conceive an integral nervousness regulator that controls the nervousness in the four defined phases. Finally, a balance between switching frequency and nervousness behaviour need to be addressed.

References

1. Barbosa, J., Leitão, P., Adam, E., Trentesaux, D.: Nervousness in dynamic self-organized holonic multi-agent systems. In: Highlights on Practical Applications of Agents and Multi-Agent Systems, pp. 9–17. Springer, Berlin, Heidelberg (2012)
2. Blackburn, J.D., Kropp, D.H., Millen, R.A.: A comparison of strategies to dampen nervousness in MRP systems. Manage. Sci. **32**(4), 413–429 (1986)
3. Hadeli, K., Valckenaers, P., Verstraete, P., Germain, B.S., Brussel, H.V.: A study of system nervousness in multi-agent manufacturing control system. In: Brueckner, S., Serugendo, G.D. M., Hales, D., Zambonelli, F. (eds.) Engineering Self-Organising Systems, Lecture Notes in Computer Science, vol. 3910, pp. 232–243. Springer, Heidelberg (2005)
4. Heisig, G.: Planning Stability in Material Requirements Planning Systems, vol. 515, Springer Science and Business Media, Berlin Heidelberg (2012)
5. Herrera, C.: Cadre générique de planification logistique dans un contexte de décisions centralisées et distribuées (Doctoral dissertation, Université Henri Poincaré-Nancy I) (2011)
6. Jimenez, J.F., Bekrar, A., Trentesaux, D., Rey, G.Z., Leitao, P.: Governance mechanism in control architectures for flexible manufacturing systems. IFAC-PapersOnLine **48**(3), 1093–1098 (2015)
7. Leitão, P., Restivo, F.: ADACOR: a holonic architecture for agile and adaptive manufacturing control. Comput. Ind. **57**(2), 121–130 (2006)
8. Leitão, P.: Agent-based distributed manufacturing control: a state-of-the-art survey. Eng. Appl. Artif. Intell. **22**(7), 979–991 (2009)
9. Minifie, J.R., Davis, R.A.: Interaction effects on MRP nervousness. Int. J. Prod. Res. **28**, 173–183 (1990)
10. Novas, J.M., Van Belle, J., Saint Germain, B., Valckenaers, P.: A collaborative framework between a scheduling system and a holonic manufacturing execution system. In: Service Orientation in Holonic and Multi Agent Manufacturing and Robotics. Studies in Computational Intelligence, pp. 3–17. Springer, Berlin (2013)
11. Onori, M., Barata, J., Frei, R.: Evolvable assembly systems basic principles. In: Information Technology for Balanced Manufacturing Systems, pp. 317–328, Springer, US (2006)
12. Pach, C., Bekrar, A., Zbib, N., Sallez, Y., Trentesaux, D.: An effective potential field approach to FMS holonic heterarchical control. Control Eng. Pract. **20**(12), 1293–1309 (2011)
13. Steele, D.C.: The nervous MRP system: how to do battle. Prod. Inventory Manage. **16**(4), 83–89 (1975)

14. Trentesaux, D., Pach, C., Bekrar, A., Sallez, Y., Berger, T., Bonte, T., Leitão, P., Barbosa, J.: Benchmarking flexible job-shop scheduling and control systems. Control Eng. Pract. **21**(9), 1204–1225 (2013)
15. Valckenaers, P., Verstraete, P., Saint Germain, B., Van Brussel, H.: A study of system nervousness in multi-agent manufacturing control system. In: Engineering Self-Organising Systems, pp. 232–243. Springer, Berlin Heidelberg (2006)
16. Verstraete, P., Saint Germain, B., Valckenaers, P., Van Brussel, H., Belle, J., Hadeli, H.: Engineering manufacturing control systems using PROSA and delegate MAS. Int. J. Agent-Oriented Softw. Eng. **2**(1), 62–89 (2008)
17. Wilensky, U.: NetLogo. Center for Connected Learning and Computer-Based Modeling. Northwestern University, Evanston, IL (1999). http://ccl.northwestern.edu/netlogo/

Part V
Service Oriented Enterprise Management and Control

Automation Services Orchestration with Function Blocks: Web-Service Implementation and Performance Evaluation

Evgenii Demin, Victor Dubinin, Sandeep Patil and Valeriy Vyatkin

Abstract This paper presents service-oriented implementation of distributed automation systems and the results of a practical performance measurement of Web-services deployed on different platforms. In the experiments we used a technique that allows one to separate the characteristics of the Web-service, such as the delays introduced by the medium of communication. It is shown that the technology development and deployment of Web-services significantly affect their performance.

Keywords Service-oriented architecture · Web service · Cloud · Pick-and-place manipulator · Web servers · Function blocks

1 Introduction

Industrial application of the Internet of Things (IoT) architecture implies embedding intelligence and communication capabilities to machines and parts thereof. Service Oriented Architecture (SOA) [1] initially developed for general purpose computing, is becoming increasingly popular in industrial automation. In the SOA way of thinking, functionalities are encapsulated into services. Services are communicating

E. Demin · S. Patil (✉) · V. Vyatkin
Luleå University of Technology, Luleå, Sweden
e-mail: sandeep.s.patil@ieee.org

E. Demin
e-mail: evgenii.demin@ltu.se

V. Vyatkin
e-mail: vyatkin@ieee.org

E. Demin · V. Dubinin
Penza State University, Penza, Russia
e-mail: victor_n_dubinin@yahoo.com

V. Vyatkin
Aalto University, Helsinki, Finland

© Springer International Publishing Switzerland 2016
T. Borangiu et al. (eds.), *Service Orientation in Holonic and Multi-Agent Manufacturing*, Studies in Computational Intelligence 640,
DOI 10.1007/978-3-319-30337-6_20

213

with others using the message passing mechanism. A service sends a request message, another service receives the message, executes the service invoked and sends a response message if needed.

Cloud computing, that is getting increasingly popular in various IT applications, can provide a very useful complement to IoT and SOA. The use of Cloud-deployed web-services in combination with embedded intelligence is being widely investigated for industrial automation applications. An example of such research activity is Arrowhead project sponsored by ARTEMIS.[1]

According to [2] "cloud computing is a modern model for ubiquitous and convenient on-demand access to a common pool of configurable remote computing and software resources and storage devices, which can be promptly provided and released with minimal operating costs and/or calls to the provider".

Cloud computing is applied in various domains, from research and media to the mail services, corporate computer systems and electronic commerce. Consumers of cloud computing can greatly reduce the cost of maintaining their own information technology infrastructure, and dynamically respond to changing computing needs in peak time periods, using the property of elasticity of cloud services.

In the development of cloud-based systems, a wide range of programming languages, libraries and technology frameworks can be used, which determine the effectiveness of the software functioning. The task of choosing the adequate development tools is an important stage of the software lifecycle. Thus, the urgent task is to study and perform comparative analysis of software applications productivity, depending on the development tools, as well as its deployment environment. Given the wide spread of distributed information systems such research is of particular interest for the Web-services—applications based on service-oriented model of interaction between providers and consumers of information services. Some work [3, 4] present a model that helps selecting best end-point for a service and this is particularly applicable in a distributed system that we are interested in and what is briefly presented in this paper.

The aim of the paper is to investigate the performance of web services developed to complement the embedded mechatronic intelligence using different development languages and deployment tools, and identify the various components of the total service time: the transmission delay/service request, and the processing time of the request and response formed by Web services.

The rest of the paper is structured as follows: Sect. 2 details the IEC 61499 function block implementation of the services, Sect. 3 presents the case study considered to demonstrate our approach, Sect. 4 presents the method of our testing approach and finally Sect. 5 presents the results and evaluation.

---

[1]www.arrowhead.eu.

2   Function Block Implementation of Services

Given the obvious lack of system-level architecture for SOA-based automation systems, some authors of this paper proposed for this purpose in [5–7] the use of IEC 61499 distributed automation reference architecture.

A family of reference examples with increasing level of complexity was considered. The first one in Fig. 1a consists of just one linear motion pusher. Once a workpiece is placed in front of the pusher (that is detected by WPS sensor), the desired service of this system is to push the workpiece to the destination sink and retract the pusher to the initial state. The figure also shows the hardware architecture of this system that fits to the Internet of Things vision: here all sensor and actuator devices are equipped with embedded microcontrollers and network interfaces.

A design environment based on such system-level architecture is required for both development and debugging. In that proposal, function block diagrams implement service diagram of SOA-based systems. Each function block refers to a service in the SOA. Connections between function blocks are reflected as messages between services. Thus, the entire function block network can be recognized as a service diagram as shown in the Fig. 1b. Each event connection inside function block design is considered as a SOAP message type. Data connections associated with this event connection are placed in the SOAP message content. SOAP messages allow a two-way communication: request and response. Similar to the functions in programming languages, a service provider can provide response messages back to the service requestor after execution is completed. Figure 2 shows this message interaction.

Section 3 presents a complete and a more complex use case and the rest of the paper deals with this case study.

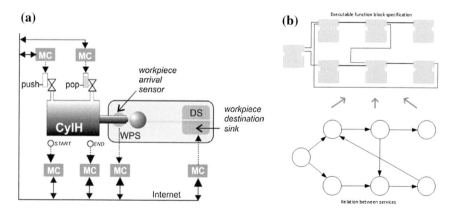

Fig. 1 **a** Workpiece transfer system with one linear motion pusher. **b** A function block application generated to implement requirements specified in the form of services

Fig. 2 SOAP response
message support in IEC
61499

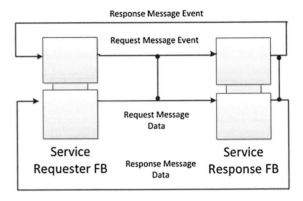

3   Case Study: Pick and Place Manipulator

This study was performed using a simulated model of pick and place
(PnP) manipulator presented in Fig. 3. The manipulator, consisting of two axes of
pneumatic cylinders and a suction device, performs the function of moving items
(work pieces) from one place to another. This manipulator has a fully decentralized
control based on collaboration of controllers embedded into each cylinder. This
architecture allows Plug and Play composition of mechatronic devices. One
approach to totally decentralised manipulator control implemented using the IEC
61499 standard is described in detail in [8, 9].

The PnP-manipulator is an automated system consisting of intelligent mechatronic components, e.g. pneumatic cylinders. Several configurations of the
manipulator are described in [10, 11]. Here we use a configuration with 6 cylinders
(3 vertical and 3 horizontal). Each cylinder can be moved in and out by the

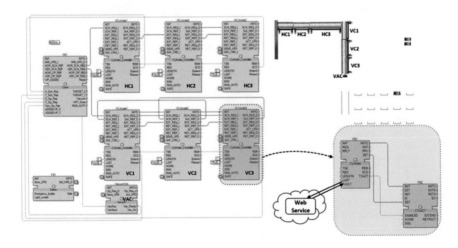

Fig. 3 Interaction scheme of the PnP-manipulator components

appropriate control signals issued by its embedded controller. A decentralized system is used to plan the motion of the PnP-manipulator cylinders, i.e., the combination of cylinders required for part delivery from a designated place. The logic of the PnP-manipulator was designed and implemented using IEC 61499 in the nxtStudio 2.1 development environment. This platform is used for the design and deployment of applications based on IEC 61499.

The intelligent mechatronic architecture demonstrated in the PnP-manipulator enables use any of mechatronic modules and extensions. This offers significant business benefits in terms of flexibility and maintainability of products.

For the experiment we have chosen a configuration of the PnP-manipulator with a separate module that implements Web-based service. This configuration has been developed and described previously in [12]. In this configuration, the planning of the cylinder movement (namely the scheduling of which cylinders participate in the given job of picking and placing of the workpiece) is performed using an external Web-Service. The interaction scheme of the PnP-manipulator components is given in Fig. 3.

4 Testing Methods

In order to study the effect of different platforms deploying Web-services on performances, we used a technique that allows separation of the transmission delay request and results from service request' processing time for each component of the system [13]. For this operation, four timestamps were recorded: (1) T_1—the time of sending a request to the Web-service by the client embedded in mechatronic component; (2) T_2—the time of receipt of the client's request by the Web-service; (3) T_3—the time of sending of response by Web-service; (4) T_4—the time when the response is received.

Fixing these timestamps allows one to estimate the following performance of the Web-service:

1. Total service time (T_{RT})—the time difference between sending the request by the client and the time of receipt of the reply from the Web-service:

$$T_{RT} = T_4 - T_1; \qquad (1)$$

2. Service delay by the web-service (T_{RPT})—time web-service takes to process the request:

$$T_{RPT} = T_3 - T_2; \qquad (2)$$

3. Transmission delay introduced by remote execution of the service (T_{RTT})—the time spent on data transfer of request/response between t client and Web-based service:

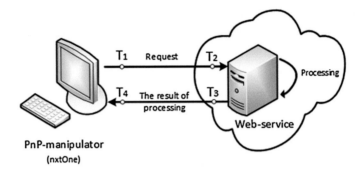

Fig. 4 Measuring the duration at different points in a single Web service request/response

$$T_{RTT} = T_{RT} - T_{RPT} = (T_4 - T_1) - (T_3 - T_2). \tag{3}$$

To analyse the changes in the performance characteristics of deployment platforms, a computer with the characteristics: CPU: Intel Core i5-2500 3.30 GHz, RAM: 8.0 GB, OS: Windows 8.1 was used when establishing and running the Web-server and Web-deployed service for planning and execution of cylinders' motion [14]. The duration of system testing was one hour. The interval between requests was 2 min [15] (Fig. 4).

5 Performance Evaluation

The experimental results show that the time of service request of the Web service greatly depends on the implementation technology and platform on which it is deployed. Based on the data obtained during the experiments, and using statistical analysis, one can formulate the following practical conclusions about the performance of deployment platforms from different manufacturers:

1. Despite the fact that Oracle and IBM products provide more flexibility when designing Web services, the performance of real-time applications is reduced.
2. When deploying Web service with Microsoft Internet Information Services (MS IIS) the real productivity is twice the one of IBM WebSphere Application Server.
3. Borland's Web-platform has shown significant volatility, in contrast with Microsoft's one. At the same time there is a steady, albeit small deviation of the average processing time for T_{RPT} for a Web-service deployed on a local server.
4. All Web-platforms use different technology to optimize Web-services.
5. Network latency has a significant effect on the performance characteristics of Web services. In addition, the timing of network connections is characterized by significant levels of volatility compared to the time of service request for a Web-service.

6. The instability of the network environment interaction of the consumer and the provider of Web services has significant impact on the uncertainty of performance characteristics. The uncertainty can be characterized by the coefficient of variation, which determines the ratio of the standard deviation and the expectation of service time. In some cases, this option is too high, which indicates substantial uncertainty of Web-platform performance characteristics measured experimentally.

The above chart shows the results of testing of Web based platforms from different vendors. Figure 5 shows the change of service time Web-service statistics. Figure 6 shows the statistics of change in network latency (Table 1).

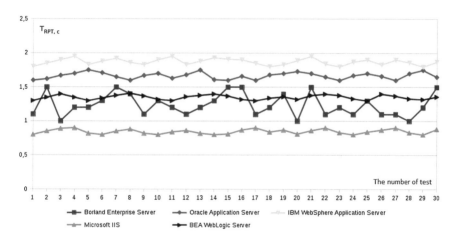

Fig. 5 Statistics change the time of service request Web-service T_{RPT}

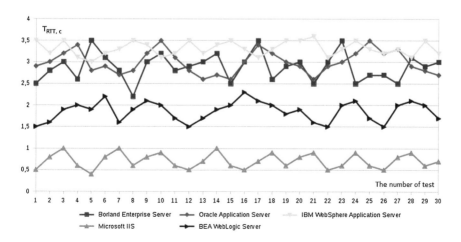

Fig. 6 Statistics changing of network delay T_{RTT}

Table 1 Statistics estimating the total service time TRT

	Borland enterprise server	Oracle application server	IBM WebSphere application server	Microsoft IIS	BEA WebLogic server
The minimum value, s	1.0	1.6	1.8	0.8	1.3
The maximum value, s	1.50	1.75	1.95	0.90	1.41
Expected value	1.24	1.66	1.87	0.84	1.35
Standard deviation	0.1673	0.0484	0.0460	0.0352	0.0351
Coefficient of variation %	0.6548	0.0408	0.0328	0.0425	0.0264

6 Conclusion

The experimental performance evaluation of Web services is important for selecting a platform for a particular application. Platform manufacturers do not provide such a comparative assessment of the performance of the development platform. Therefore, the end user cannot predict the cost of the equipment required for the deployment of distributed or cloud computing applications. As a general limitation of this study, it should be noted that this method of testing does not account for the impact of delays in the communication environment of Web services and Cloud applications themselves.

The difference in the implementation of modern technologies of Web-based applications, as well as the instability of the characteristics of the Internet environment interaction have a significant impact on the performance of Web-based software and cause a significant degree of uncertainty in their practical utility. At the same time, experimental studies such as those described in this paper are important because they allow developers and users of Web-based applications choosing between information technology implementation and deployment methods, as well as the prediction of non-functional characteristics.

Acknowledgments This work was partially supported by the program "Fundamental research and exploratory research involving young researchers" (2015–2017) of the Russian Science Foundation (project number 15-11-10010), and by Luleå Tekniska Universitet through the grants 381119, 381940 and 381121.

References

1. Erl, T.: Service-Oriented Architecture: Concepts, Technology, and Design. Prentice Hall PTR, Upper Saddle River (2005)
2. Jadeja, Y., Modi, K.: Cloud computing—concepts, architecture and challenges. In: 2012 International Conference on Computing, Electronics and Electrical Technologies (ICCEET), pp. 877–880 (2012)

3. Serbănescu, V.N., Pop, F., Cristea, V., Achim, O.M.: Web services allocation guided by reputation in distributed SOA-based environments. In 11th International Symposium on Parallel and Distributed Computing (ISPDC), pp. 127–134 (2012)
4. Achim, O.M., Pop, F., Cristea, V.: Reputation based selection for services in cloud environments. In 14th International Conference on Network-Based Information Systems (NBiS), pp. 268–273 (2011)
5. Dai, W., Vyatkin, V., Christensen, J.H.: The application of service-oriented architectures in distributed automation systems. In 2014 IEEE International Conference on Robotics and Automation (ICRA), pp. 252–257 (2014)
6. Dai, W., Christensen, J.H., Vyatkin, V., Dubinin, V.: Function block implementation of service oriented architecture: Case study. In: 12th IEEE International Conference on Industrial Informatics (INDIN), pp. 112–117 (2014)
7. Dai, W., Riliskis, L., Vyatkin, V., Osipov, E., Delsing, J.: A configurable cloud-based testing infrastructure for interoperable distributed automation systems. In IEEE International Conference on Industrial Electronics IECON'14, Dallas (2014)
8. Vyatkin, V.: IEC 61499 as enabler of distributed and intelligent automation: State-of-the-art review. IEEE Trans. Industr. Inf. 7(4), 768–781 (2011)
9. Sorouri, M., Patil, S., Vyatkin, V., Salcic, Z.: Software composition and distributed operation scheduling in modular automated machines. In IEEE Transactions on Industrial Informatics, vol. 11, pp. 865–878 (2015)
10. Patil, S., Yan, J., Vyatkin, V., Pang, C.: On composition of mechatronic components enabled by interoperability and portability provisions of IEC 61499: A case study. In 18th IEEE Conference on Emerging Technologies and Factory Automation (ETFA), pp. 1–4 (2013)
11. Vyatkin, V.: Intelligent mechatronic components: Control system engineering using an open distributed architecture. In IEEE Conference on Emerging Technologies and Factory Automation, Proceedings ETFA '03, vol. 2, pp. 277–284 (2003)
12. Demin, E., Patil, S., Dubinin, V., Vyatkin, V.: IEC 61499 Distributed control enhanced with cloud-based web-services. In IEEE Conference on Industrial Electronics and Applications, Auckland, New Zealand (2015)
13. Feng, L., Gesan, W., Li, L. Wu, C.: Web service for distributed communication systems. In IEEE International Conference on Service Operations and Logistics, and Informatics (SOLI '06), pp. 1030–1035 (2006)
14. Cheung, L., Golubchik, L., Fei, S.: A study of web services performance prediction: A client's perspective. In 19th IEEE International Symposium on Modeling, Analysis and Simulation of Computer and Telecommunication Systems (MASCOTS), pp. 75–84
15. Velkoski, G., Simjanoska, M., Ristov, S., Gusev, M.: CPU utilization in a multitenant cloud. In EUROCON, IEEE, pp. 242–249

IoT Visibility Software Architecture to Provide Smart Workforce Allocation

Pablo García Ansola, Andrés García and Javier de las Morenas

Abstract In manufacturing and logistics companies there are many processes and services that cannot be fully automated and the integration with workforce is the key to provide better results. One example is Airport Ground handling operations where agents, operators, drivers or aircraft crews need to generate and feed information from other processes and events in order to provide better schedules. This work uses manufacturing and Internet of Things (IoT) concepts to design software architecture to generate an uncoupled workforce information feedback for current agent-based decision-making frameworks. In the case at hand, the architecture is implemented in a cloud-based commercial solution called "aTurnos", which has already been deployed by different companies to schedule working shifts for over 25,000 employees. The handling company being analysed in this paper requires a dynamic allocation of employees and tasks with updated field information about the status of workers in real time.

Keywords Smart and digital enterprise · Workforce management · Service oriented agents and MAS (SoMAS) · Decision support system (DSS)

P.G. Ansola (✉) · J. de las Morenas
AutoLog Group, Mining and Industrial Engineering School of Almadén,
University of Castilla-La Mancha, Almadén, Spain
e-mail: pablo.garcia@uclm.es

J. de las Morenas
e-mail: javier.delasmorenas@uclm.es

A. García
AutoLog Group, School of Industrial Engineering, University of Castilla-La Mancha,
Ciudad Real, Spain
e-mail: andres.garcia@uclm.es

© Springer International Publishing Switzerland 2016
T. Borangiu et al. (eds.), *Service Orientation in Holonic and Multi-Agent Manufacturing*, Studies in Computational Intelligence 640,
DOI 10.1007/978-3-319-30337-6_21

223

1 Introduction

Given the actual high level of operational disaggregation in worldwide companies, decision-making units need to drive their business decisions to a common framework where all their parts support the global objectives of the company. This necessity of integration brings forward concepts such as *alignment*, which has become particularly important in the field of business integration and interoperability [1, 2]. Specifically, workforce management requires new management tools in complex manufacturing/transport/logistics infrastructures as it is nowadays experiencing an increase in complexity given the new requirements for a more flexible environment, with more workers, higher turnover and more dynamic, uncertain and complex assignments [3]. A flexible workforce management implies an annual or seasonal planning for the definition of contracts and union validations, while at the same time being able to perform a comprehensive and dynamic reallocation of tasks for daily operations. In the airport ground handling scenario, the ground-handling agents have to modify their location, role and timing attending to on-going changes in the environment, therefore requiring of powerful monitoring and allocation systems. Furthermore, all activities in this sector involve the cooperation of international regulation bodies, private companies, clients, subcontractors and operators. All these parties get involved in the same decision process but with objectives that can be shared or specific along a decision-making process that becomes specially hard to align when subject to a variety of disturbances [4]. Therefore the resulting systems must be capable of providing a high level of intelligence with efficient data capture, analysis and decision-making capabilities which are related to the real time location of human resources, products and processes. Hence, the first step for the deployment of such a system is to provide flexible visibility frameworks, which have to comply with legal restrictions referring to privacy, safety and quality of service [5]. For example, Data Protection Agencies [6] have to certify that the gathering of information is limited to that required for process improvement. Even so, the management of this kind of information needs always to be agreed with workers unions. Thus, the suitability of these systems needs to be clearly motivated in terms of safety, security and quality. At this point, IoT is developing concepts and technologies that are dealing with the real time information generated by the disturbances that are common in these environments [7].

Specifically, the manufacturing research community defines the "Internet of Things" as the tool to generate new mechanisms based on a virtualization of physical resources to link the real world and the computers [8]. The "Internet of Things" adds another data dimension, as it allows the physical world, things and places, to generate data automatically by sensing the physical world [7]. At this point, the main objective of "IoT" addressed by this paper is to automatically capture the information on human resources when new events occur at shop floor level even if a lack of dynamic integration with decision-making process is nowadays usual [9].

2 Background Considerations

There are many planning algorithms that solve problems of workforce allocation such as the scheduling of shifts for nurses in a hospital. These problems require the implementation of relatively complex programming algorithms such as backtracking with high computational requirements. For the application at hand, some authors asserted that optimal solutions are not required; good ones are enough as is the case for the results obtained by heuristic or voracious algorithms [9, 10]. The lack of standards for the data provided to algorithms is another problem of planning systems, which thus require custom developments that increase costs. One of the most important weaknesses in online workforce management is the lack of information about the real evolution that is taking place in the environment. Deviations in allocations are not automatically handled by the planning system, but manually by supervisors or team leaders in the workplace (e.g. a factory). To make this possible it is necessary to design interfaces capable to capture real and reliable information and to make it available to all parties by publishing it in a standard format.

For example, when addressing the problem with a vision related to manufacturing environments, the first step is to identify the point within the scheduling process where integration of physical data and decision-making need to be implemented; this corresponds to the point in which the information from the environment needs to be fed. Firstly, at manufacturing cell level, the system executes the low level actions through machine control subject to its specific constraints. Above cell level, shop floor is the lowest level that introduces flexibility requirements; on this level the tasks/orders are broken into single instructions executed downstream by the cells that directly manipulate products and resources. This level supposes the first cooperation between elements. It schedules the instructions and controls the operational disturbances between operational cells. Shop floor control is often referred as Manufacturing Execution Systems (MES) [11]. The Decision Support Systems (DSS) require information coming from both adjacent levels: the enterprise level (information systems) and the cell level (physical resources). More specifically, in the manufacturing/production control proposed by McFarlane et al. [12], the product scheduling is an interface between the resource status and the order status. In a workforce shift-allocation process, the visibility requirements are the same, those of checking the status of workers while being fed constraint information from the upper levels related do factory management. Therefore, the visibility framework used in the definition of the main requirements and constraints is crucial in workforce management. To set up this visibility framework it is necessary to start by selecting technologies for the tracking of workers attending to the use cases, environment and employment laws.

Some common requirements for a workforce visibility framework are:

- The definition of the workforce area as an indoor/outdoor environment. The tracking is forbidden out of the work environment.
- Employee monitoring cannot be continuous but limited to specific positions, crossing points and doors.

- The system must be autonomous and easy to deploy without using existing equipment or infrastructures. It may have to overcome coverage problems. As would be the case in a demo that can gather more than 5000 users/employees.

Even as that is not the focus of this paper a short overview of the hardware is nevertheless required to understand the software architecture. The present approach uses regular smartphones that identify Bluetooth points based on the iBeacons protocol, which is a commercial distribution of Bluetooth 4.0 from Apple. These iBeacons were initially designed for commercial/marketing issues and are well known by the industry but, as in this case, they may be used for many other purposes. In the case of workforce management, the Bluetooth points or beacons are distributed in the scenario and broadcasting their unique ID with a configurable coverage that goes from 1 to 60 m. These beacons define locations making possible to cover big spaces by providing specific identifications at the defined strategic points. The smartphones carried by employees send the locations IDs to the cloud when they are in the proximity of a beacon through a process called "check". The next figure details the setting up of this network or mesh. The communication between iBeacons and smartphones uses Bluetooth, and the smartphones send their ID to the cloud using the company network (3G or WIFI) (Fig. 1).

3 Software Architecture Incorporating EPCIS

The proposed software architecture can directly benefit from visibility frameworks through a software interface such as standard Electronic Product Code Information Services (EPCIS), a well-known software architecture for visibility frameworks used in manufacturing and logistics. The EPCIS specification helps defining this dynamic interface by using the existing services, which automatically publish real-time information coming from the plant and its circumstances, reporting processed information and abstracting upper IT levels. In the proposed system, EPC subscriptions connect the human resources with the internal reasoning of the affected agents. The beliefs of these agents are updated following the standard EPCIS XML specification [3]. The proposed architecture has been implemented in a cloud-based commercial solution called "aTurnos", which has already been commercially deployed by different companies to schedule working shifts for over 25,000 employees. This solution requires to identify what services will be needed to test the extrapolation of EPCIS to workforce management. The services implemented in the aTurnos Web services that are already available to its client companies through their App are:

- setCheck(EPC-Package), receives the new location of the employees based on the EPCIS standard as detailed below.
- setNewLocation(Location), receives a new location for the employee being identified.

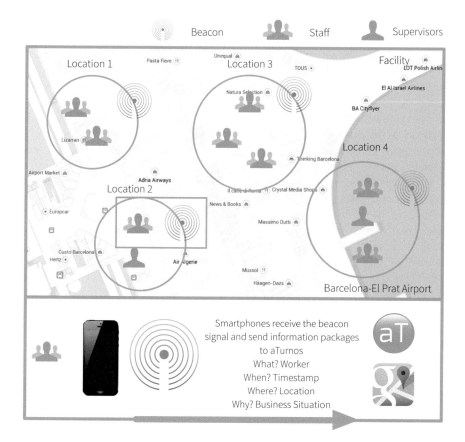

Fig. 1 Hardware design

- getLocationList(), sends to the employees the location list defined in the workspace.
- getTaskList(), pushes the updated task list to the employees when disturbances occur.

The employees' App in their smartphones use the setCheck() service to send a customized package on JavaScript Object Notation (JSON). This service is based on the EPCIS specification of What, Where, When and Why and gets triggered by the identification of an iBeacon during operation. Specifically, the definition of this package is:

- What? The resource—employee in this case—that is logged in the App.
- When? The timestamp when the identification occurs.
- Where? The location (ID of beacon) where the employee is situated.
- Why? The work process being carried out by the worker that is related to the on-going tasks (i.e. the reason for the resource to be there at that time).

Looking at the full cycle, the employees close to the beacons send a check to the server in a JSON package. Then, the employee avatar implemented in an agent is subscribed to the events of this employee using the EPC products subscription process. Based on this information, the agent-based system is capable to reschedule the plan, thus allocating new tasks to the employees. Finally, the tasks service pushes the information to the smartphones of the involved employees. In order for the system to being able to perform all these operations in an efficient manner, it is necessary to uncouple the problem while maintaining the cohesion between checks and the agent information used in the decision-making process. This work proposes the definition of states based on these checks (What? Where? When? and Why?).

The checking process allows identifying the status of employees in an uncoupled way. This simplifies implementation by tackling the problem of the excessive dependence on up-to-date field information, which is usually the draw-back of MAS [11, 13]. As discussed above, the status of an employee is defined by the "What?" (Product identification), "Where?" (Read point), "When?" (Timestamp) and "Why?" (Business step) information. In Eqs. 1–3 "p" represents the employees which have a set of Reader Points (R) (Eq. 1) and Business Steps (B) (Eq. 2). In an airport ground-handling scenario, if an operator is busy (why?) in a specific gate (where?) it defines a state. As the number of business steps and reading points grow, the precision during the decision-making gets bigger because of the corresponding increase in the number of states. The definition of states is flexible, so new read points and business steps can be added during operations.

$$R = (r_1, r_2, \ldots, r_n), \quad R \in \mathcal{M}_{1xn} \tag{1}$$

$$B = (b_1, b_2, \ldots, b_m), \quad B \in \mathcal{M}_{1xm} \tag{2}$$

$$S = \begin{bmatrix} s_{00} & \cdots & s_{0n} \\ \vdots & \ddots & \vdots \\ s_{m0} & \cdots & s_{mn} \end{bmatrix}, \quad s_{ij} = r_i x b_j, \quad S \in \mathcal{M}_{mxn} \tag{3}$$

Once the employee states in the schedule are identified, it is possible to define the transactions between the current state (s_{ij}) and the desirable state (s'_{ij}) by using a simple inference of B and R. Any new event coming from the checks triggers a reschedule based on the modification of the state. There is an event subscription between the checks service and the agent-based avatar in aTurnos.

Implementing the "aTurnos" solution, the initial phase that needs to be addressed must situate every location in a Map (the actual version uses Google Maps). This will allow aTurnos to check possible delays and trigger rescheduling processes based on distances of the resources (workers) to the required operation points and timings. The example in Fig. 1 shows the Airport of Barcelona, where 4 iBeacons have been situated in different points at terminals and aTurnos can calculate the shift times between locations at the Google Maps API. This was an efficient (i.e. fast) way to generate information about the real environment with third party software in a dynamic interface.

IBeacons stands as an appropriate solution because it makes possible increasing the precision as required by adding beacons to the map. When new beacons are detected they can be submitted to the server as new locations based on the GPS location of the employee that discovers the new iBeacon. At this point, any employee within the App can generate new visibility points during operations that can be supported by the company or even by third parties.

4 Establishing a Test Bench Scenario

Several ratios have to be introduced in order to ascertain the performance of the system: control over the deviations in supervising tasks, tracking positions, time lost in communication within the team or even the validation of the quality of service. This example takes the simple approach of using "Dead Time" (DTe) defined as the time elapsed between the required starting time of a shift and the time at which it really starts (4). Therefore, DTe can be calculated as:

$$DTe = Tnt + Tmove \tag{4}$$

where Tnt stands for the time at which employees have not yet been informed about an assigned task and Tmove is the time that an employee needs to move to the new location (5). Tnt includes the time required for the supervisor to identify the deviation over the current plan, the time to plan and the time to communicate the new instructions to the team.

$$Tnt = Identification + Plan + Comunication \tag{5}$$

Figure 2 illustrates statistical data that shows how traditional decision-making is adapted to flexible requirements in complex operations. This information is generated by aTurnos based on their current customers from the logistics sector. The red bars show the current number of employees (coverage) at every hour and the

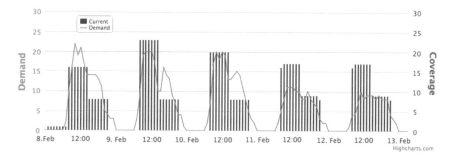

Fig. 2 Current decision-making behaviour

green line defines the real process demand per hour. The demand is obtained after the event has occurred but supervisors have access to historical data and demand forecasting. The first day, existing employees do no cover the real needs. The second day, the supervision contracts more workers but they are concentrated in the morning shift with a shortage remaining for the afternoon. The third day, this deviation persists because the information used by the supervisor to define the new plan is not accurate enough. For this case, the coverage does not fit the requirements until the 4th day, even though for two days there were more workers allocated to the morning shift than really needed.

The proposed system would make possible reducing this waste time by improving the scheduling process using cloud-based solutions. Based on the Tnt, the reduction of waste time can be considered in three aspects:

- The automatic identification of disturbances; when there are many tasks and resources it is hard to identify the disturbances and delays.
- The new plans are directly generated by the agent-based system every time these disturbances occur. The managers can access the real time information in the cloud through their mobile terminals during operations and validate or re define processes.
- The fast communication to the employees through the smartphone with the new plan. If this process is manually, the supervisor needs to communicate every new instruction.

5 Conclusions

This work presents a software architecture suitable for dynamically rescheduling work force tasks during operations. The proposed system takes advantage of many of the concepts, methods and technologies that are commonly used by the manufacturing research community. Specifically, the architecture takes advantage of a customized EPCIS in an on-cloud solution that provides cohesion and uncoupling from the agent-based decision makers. Thus, dynamic re-scheduling is performed automatically and the information duly updated to the involved employees. This allows a real time workforce management with control of shifts, working hours per employee, costs and productivity. Real results need to be addressed in a bigger variety of scenarios; however, initial applications have shown a remarkable improvement in workforce management. This has been considered by aTurnos and by its clients as a feasible software solution in terms of expenses, quality, transparency and service. Future research will focus on the integration of product processes with these workforce management solutions, which is bound to provide better decision-making solutions in global companies where processes involve workers together with other resources.

References

1. Zhang, P., et al.: The Influence of Industries and Practice on the IS Field: A Recount of History in Twentieth Americas Conference on Information Systems, Savannah (2014)
2. Luftman, J.: Strategic alignment maturity. Handbook on Business Process Management 2, pp. 5–43. Springer, Berlin, Heidelberg (2015)
3. García Ansola, P., et al.: Distributed decision support system for airport ground handling management using WSN and MAS. Eng. Appl. Artif. Intell. **25**(3), 544–553 (2012)
4. Ashford, N., Stanton, H., Moore, C.: Airport Operations. McGraw-Hill Professional, New York (1998)
5. Flynn, B.B., Schroeder, R.G., Sakakibara, S.: The impact of quality management practices on performance and competitive advantage. Decis. Sci. **26**(5), 659–691 (1995)
6. Liu, L.: From data privacy to location privacy: Models and algorithms. In Proceedings of the 33rd International Conference on Very Large Data Bases (VLDB) Endowment (2007)
7. López, T.S., et al.: Adding sense to the internet of things. Pers. Ubiquit. Comput. **16**(3), 291–308 (2012)
8. Da Xu, L., He, W., Li, S.: Internet of things in industries: A survey. IEEE Trans. Industr. Inf. **10**(4), 2233–2243 (2014)
9. Colombo, A.W., et al.: Industrial cloud-based cyber-physical systems, The IMC-AESOP Approach (2014)
10. De Weerdt, M., Clement, B.: Introduction to planning in multiagent systems. Multiagent Grid Syst. **5**(4), 345–355 (2009)
11. Leitão, P.: Agent-based distributed manufacturing control: A state-of-the-art survey. Eng. Appl. Artif. Intell. **22**(7), 979–991 (2009)
12. McFarlane, D., et al.: Intelligent products in the supply chain—10 years on. Service Orientation in Holonic and Multi Agent Manufacturing and Robotics, pp. 103–117. Springer, Berlin, Heidelberg (2013)
13. Archimede, B., et al.: Towards a distributed multi-agent framework for shared resources scheduling. J. Intell. Manuf. **25**(5), 1077–1087 (2014)

Virtual Commissioning-Based Development and Implementation of a Service-Oriented Holonic Control for Retrofit Manufacturing Systems

Francisco Gamboa Quintanilla, Olivier Cardin, Anne L'Anton and Pierre Castagna

Abstract While cyber-physical systems probably represent the future of industrial systems, their development might take some time to be extensively applied in industry. This paper presents the implementation of a service-oriented holonic control on a pre-existing system. The development of the control system is based on a virtual commissioning phase, developed with a Rockwell Arena simulation model.

Keywords HMS · SoA · FMS · Emulation · Virtual commissioning

1 Introduction

One of the key objectives of current research activities worldwide is to define best practices for implementing agile manufacturing systems. One very promising trend deals with the breakthrough induced by cyber-physical systems technology [1–3], for production [2], maintenance [4, 5] or logistics issues [6] to name a few. Even if these technologies are of a great interest and innovative solutions will appear soon on the market, the fundamental change induced and the cost of system's enhancements might impose a delay of a couple of decades between their industrial maturity and their exploitation in a large scale. Based on a size 1 experimental platform, this paper intends to define a framework for implementing an agile control

F. Gamboa Quintanilla · O. Cardin (✉) · A. L'Anton · P. Castagna
LUNAM Université, IUT de Nantes—Université de Nantes, IRCCyN UMR CNRS
6597 (Institut de Recherche en Communications et Cybernétique de Nantes),
2 avenue du Prof. Jean Rouxel, Nantes 44475, Carquefou, France
e-mail: olivier.cardin@univ-nantes.fr

F. Gamboa Quintanilla
e-mail: francisco.gamboa@univ-nantes.fr

A. L'Anton
e-mail: anne.lanton@univ-nantes.fr

P. Castagna
e-mail: pierre.castagna@univ-nantes.fr

© Springer International Publishing Switzerland 2016
T. Borangiu et al. (eds.), *Service Orientation in Holonic and Multi-Agent Manufacturing*, Studies in Computational Intelligence 640,
DOI 10.1007/978-3-319-30337-6_22

on a pre-existing manufacturing system. The chosen *control is a service-oriented holonic manufacturing system* control (SoHMS) [7, 8] developed on a multi-agent platform. To reduce the cost of development, a virtual-commissioning based approach is presented, with all the constraints it implies on the control architecture.

Section 2 describes the system under study. Section 2.4 presents the current and targeted control systems. Section 3 describes the virtual commissioning phase and finally Sect. 4 explains the integration on the real system.

2 System Description

2.1 Flexible Manufacturing System

The application of the SoHMS is made to a small production line located at the University of Nantes, France. This production line, Fig. 1, is an automated assembly line composed of three workstations and a conveyor system formed by four conveyor loops of which three serve as buffers for each of the workstations and the other, the main loop, serves for transportation between the workstations. Product goods are transported by the conveyors with pallets having an intelligence level 1 [9], containing capabilities of self-identification with RFID tags in order to allow the transport resources to direct the pallet through the conveyors diverters from one port to another.

Workstations are composed of 6-axis robotic arms, a stock of Lego® blocks, a temporary stock and a workspace location for incoming pallets, Fig. 2.

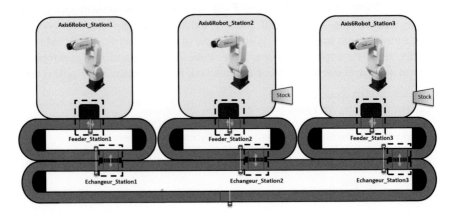

Fig. 1 Production line layout

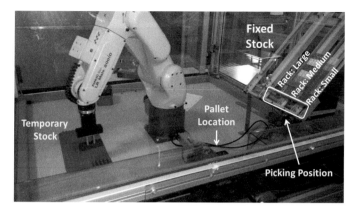

Fig. 2 Workstation layout

2.2 Products

The main function of the robotic arm is to perform pick and place assembly operations. The main task of the robot is to pick a corresponding Lego® block from the fixed or temporary stock and assemble it on the product under treatment. The fixed stock has three different racks, each rack for a different size of Lego® block. Within a rack, blocks of different colours may arrive randomly way. When a special colour is demanded and is not available in the picking position of the rack, the robot can use the temporary stock to remove blocks from the rack to make the desired colour available.

The product is a structure of Lego® blocks compiled in a specific configuration. Figure 3 illustrates a product family with two versions sharing the same product

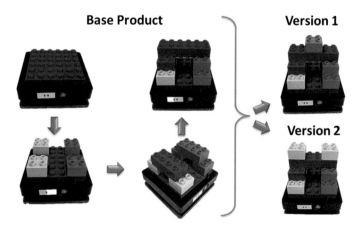

Fig. 3 Product example, with possibility of customization in 2 versions

feature, therefore all members have the same process structure up to the third level. Differentiation occurs at the fourth level where two versions can be issued.

Lego® structure is formed by three types of blocks namely; a small 2×2 block, a medium 2×4 block and a Large 2×6 block. These blocks can be assembled in any position (X, Y, Z and W axes). Added to this, each block is available in four colours: red, green, yellow and blue. Hence, there is a great flexibility to create a vast variety of structures. Customization for such product family happens at a scalable level with the choice of colour and at a modular/structural level with the choice of version. The Lego® structure results to be an ideal alternative in order to illustrate, in a very simple manner, the dependencies between the different components of the structure.

2.3 Services Definition

In the SoHMS there are the product-level services which are offered by workstations and transport resources. This service library belongs to the production line which can be viewed as a resource itself thus having a service offer of the different product families it can produce. Lego® blocks are used to represent the different manufacturing services, Fig. 4. In this way, taking the three types of blocks available, the service ontology for this application is formed by three types of services per layer. Differentiated by their size; a 2×2 block represents a service class A, a 2×4 block a service class B and a 2×6 block a service class C. This constitutes an ontology of $4 \times 3 = 12$ services types namely; A1 for a small block at level 1, B2 for a medium block at level 2, C3 for a large block at level 3, etc. Moreover, each of these services has a set of parameters. These parameters are the colour and position of the block only in x and y coordinates as the vertical position forms part of the service type definition. Other product-level services are the Transport_Pallet service and the

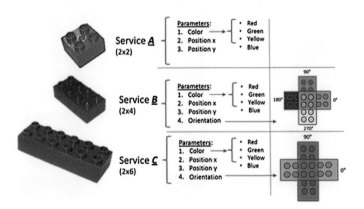

Fig. 4 Types of assembly services

Supply_Base service. The transport service has parameters: startPort and endPort while the supply service has the parameter colour of the base.

As the production line represents a flexible job-shop, service redundancy is included. Workstations 2 and 3 provide all the manufacturing services for the assembly of Lego® block of the three sizes. However, even though both workstations provide the same service types, these do not have the same capabilities at any time, considering the range of possible colours in stock for example.

2.4 The Control System

The system is equipped with control equipment settled on a TCP/IP network (Fig. 5). The robot controllers are able to communicate on Modbus TCP, same as the I/O on IP modules. On these modules, sensors and actuators (electro valves, lamps, relays) are directly wired and RFID readers are connected in RS485, communicating with serial Modbus. This architecture was originally dedicated to welcome a classical PLC. Field level orchestrations could be implemented as

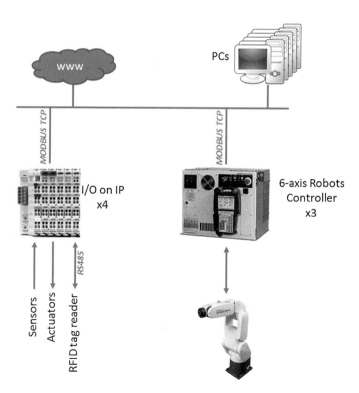

Fig. 5 Pre-existing control hardware

described in [10]. The choice made on this system is different. On the network, industrial PCs are set up and emulate a java virtual machine in order to run different programs, each having a specific function.

3 Virtual Commissioning

The emulation model was implemented with a discrete-event simulation tool, namely Rockwell Arena. Such tools are quite efficient to model activities based on queues management. In this model, queues are extensively used to model the synchronization between the events occurring on the emulation and the orders coming from SoHMS (Fig. 6).

For validation purposes, it is necessary to have a behaviour of the emulation mimicking the behaviour of low-level devices, as expressed in [11]. A TCP socket interface is therefore integrated in the emulation model, which both triggers orders and sends events information to SoHMS. It is able to understand orders such as "TRANSPORT TransporterID FROM InitialZone TO FinalZone" or "PICK RobotID Store". This emulation does not contain any intelligence, but is only able to execute a set of pre-programmed list of orders in reaction to a high-level order on the socket. The actions of the robots are transformed into delays of predetermined length.

Fig. 6 Integration of the emulation model in the control architecture

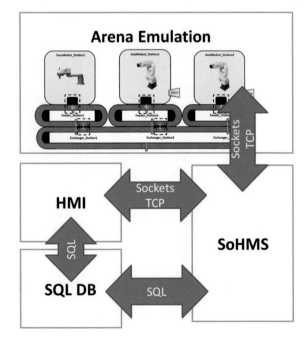

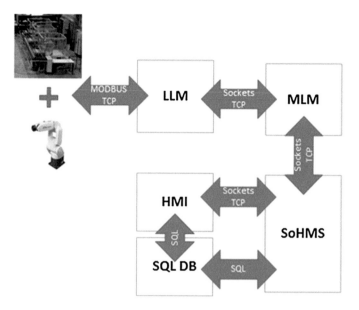

Fig. 7 Targeted control architecture

4 Integration on the Real System

The choice made on this system is to replace the PLC by ad hoc programs (Fig. 7), able to handle higher semantics than PLC do and more flexible in configuration for experimental purposes. First, a Low-Level Middleware (LLM) was created. The objective of LLM is to synchronously retrieve the state of each sensor of the system, asynchronously inform the upper layers of any change of value of the sensors and asynchronously modify the state of actuators on upper layers' order. Functionally, this is close to what OPC[1] servers do, but adapted to the hardware configuration.

Second, a Medium-Level Middleware (MLM) is in charge of aggregating the data coming form LLM for upper layers and time macro-actions requested by upper layers in high level semantics. For example, when a pick service is requested on a robot, MLM communicates to LLM all the configuration bytes to modify on the controller, waits for an acknowledgement, sends the program start order, sends an acknowledgement to upper layer that the service is running, waits for the

---

[1]www.opcfoundation.org.

acknowledgement of program end and sends an acknowledgement of service end. These functionalities are close to those of a PLC, but with higher level semantics.

Finally, the SoHMS is connected to MLM, a Human-Machine Interface and a SQL database, storing production data and results. Validated with the emulation phase, it is plug-and-play on the system.

Alternate architecture solutions can also be implemented on this system. Figure 8 shows a configuration where several SoHMS are connected to the system. This is fully transparent for the system, as MLM does not differentiate orders coming from the upper-level. Another alternate solution is presented in Fig. 9. The virtual commissioning phase is oriented toward a monolithic architecture, as the emulation model was. However, it is absolutely possible to decline the holonic architecture induced by the SoHMS to an actual decomposition with each holon having its own MLM-LLM couple. The necessary step to ease this decentralization is to make these programs dependent of configuration files, indicating the limits of the considered holon.

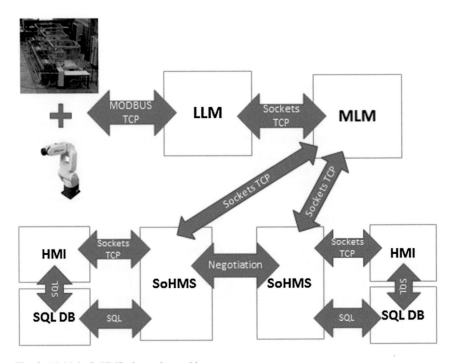

Fig. 8 Multiple SoHMS alternative architecture

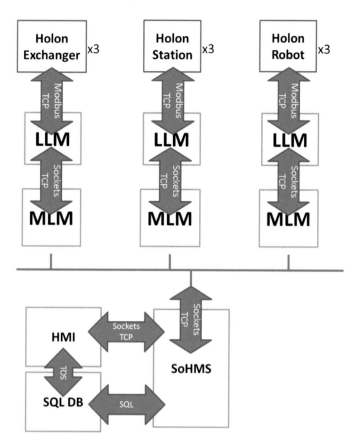

Fig. 9 Distributed alternative architecture

5   Conclusion

This study introduces a new experimental platform, built up around conveyor loops, three robotic stations and a control architecture fully programmed in ad hoc Java code. A SoHMS was implemented, thanks to a virtual commissioning phase performed via a Rockwell Arena simulation model. The next step is to generalize the programs in order to distribute the control and enhance the autonomy of holons.

References

1. Colombo, A.W., Karnouskos, S., Bangemann, T.: Towards the next generation of industrial cyber-physical systems. In Industrial Cloud-Based Cyber-Physical Systems. Springer, Berlin, pp. 1–22 (2014)
2. Lee, J., Bagheri, B., Kao, H.-A.: A cyber-physical systems architecture for industry 4.0-based manufacturing systems. Manuf. Letters **3**, 18–23 (2015)
3. Monostori, L.: Cyber-physical production systems: Roots expectations and R&D challenges. Procedia CIRP **17**, 9–13 (2014)
4. Trentesaux, D., Knothe, T., Branger, G., Fischer, K.: Planning and control of maintenance, repair and overhaul operations of a fleet of complex transportation systems: A cyber-physical system approach. In: Borangiu, T., Trentesaux, D., Thomas, A. (eds.) Service Orientation in Holonic and Multi-agent Manufacturing, pp. 175–186. Springer, Berlin (2015)
5. Zhong, H., Nof, S.Y.: The dynamic lines of collaboration model: Collaborative disruption response in cyber–physical systems. Comput. Ind. Eng. **87**, 370–382 (2015)
6. Seitz, K.-F., Nyhuis, P.: Cyber-physical production systems combined with logistic models—a learning factory concept for an improved production planning and control. Procedia CIRP **32**, 92–97 (2015)
7. Morariu, C., Morariu, O., Borangiu, T.: Customer order management in service oriented holonic manufacturing. Comput. Ind. **64**(8), 1061–1072 (2013)
8. Quintanilla, F.G., Cardin, O., Castagna, P.: Product specification for flexible workflow orchestrations in service oriented Holonic manufacturing systems. In Borangiu, T., Trentesaux, D., Thomas, A. (eds.) Service Orientation in Holonic and Multi-Agent Manufacturing and Robotics. Springer, Berlin, pp. 177–193 (2014)
9. Wong, C.Y., McFarlane, D., Ahmad Zaharudin, A., Agarwal, V.: The intelligent product driven supply chain. In 2002 IEEE International Conference on Systems, Man and Cybernetics, vol. 4, pp. 6–10 (2002)
10. Legat, C., Vogel-Heuser, B.: An orchestration engine for services-oriented field level automation software. In Borangiu, T., Thomas, A., Trentesaux, D. (eds.) Service Orientation in Holonic and Multi-Agent Manufacturing. Springer Studies in Computational Intelligence, pp. 71–80
11. Berger, T., Deneux, D., Bonte, T., Cocquebert, E., Trentesaux, D.: Arezzo-flexible manufacturing system: A generic flexible manufacturing system shop floor emulator approach for high-level control virtual commissioning. Concurrent Eng. July, 1063293X15591609 (2015)

Security Issues in Service Oriented Manufacturing Architectures with Distributed Intelligence

Cristina Morariu, Octavian Morariu and Theodor Borangiu

Abstract The paper discusses the main classes of shop floor devices relative to distributed intelligence for product-driven automation in heterarchical control. The intelligent product (IP) concept is enhanced with two additional require-ments: standard alignment and SOA capability. The paper classifies IPs from SOA integration point of view and introduces a formalized data structure in the form of a XSD schema for XML representation. We propose a security solution for service oriented manufacturing architectures (SOMA) that uses a public-key infrastructure to generate certificates and propagate trust at runtime (during product execution) for embedded devices that establish IPs on board pallets and communicate with shop floor resources. Experimental results are provided.

Keywords Manufacturing execution system · Distributed intelligence · SOMA · Intelligent product · Security · Multi-agent framework

1 Introduction

One important requirement for Manufacturing Execution Systems (MES) is the ability to react to changes: external, induced by market variability or internal, caused by unexpected events (failures) during product execution. Considering the complete manufacturing value chain, one of the most complex problems identified in practice by Wong [1] is the discrepancy between the material flow and the information flow. This discrepancy is caused usually by the fact that the product is

C. Morariu · O. Morariu · T. Borangiu (✉)
Department of Automation and Industrial Informatics, University Politehnica, Bucharest, Romania
e-mail: theodor.borangiu@cimr.pub.ro

C. Morariu
e-mail: cristina.morariu@cimr.pub.ro

O. Morariu
e-mail: octavian.morariu@cimr.pub.ro

© Springer International Publishing Switzerland 2016
T. Borangiu et al. (eds.), *Service Orientation in Holonic and Multi-Agent Manufacturing*, Studies in Computational Intelligence 640,
DOI 10.1007/978-3-319-30337-6_23

not aware of its current state, location or identity. The requirements for manufacturing agility on one hand and supply chain predictability on the other hand converged and the concept of intelligent product (IP) has emerged. McFarlane [2] presents the main characteristics of the intelligent product as the ability to monitor, assess and reason about its current and future state. At the same time, recent research offers advances in developing MES applications with Service Oriented Architecture (SOA) [3].

In this context, the intelligent product concept needs to be enhanced with two additional requirements: standard alignment and SOA capability. These requirements are vital for seamless real time integration of intelligent products with the other components (e.g. batch planning, automated execution, quality control, inventory, etc.) of the overall manufacturing system control system. On the other hand, when considering the global environment in which manufacturing companies must operate today in order to remain competitive and keep costs at minimum, the standardization problem becomes very important. Operating with proprietary information structures becomes a concern that prevents real cooperation between organizations towards a common goal. We argue that the alignment to standards will be a decisive factor for determining the manufacturing enterprise's success in the years to come [4].

Several standards have emerged in the last period supporting the advances in SOA adoption in manufacturing, some of the best known examples being: ISA 95, ISA 88, ebXML, EDDL, FDT and MIMOSA. The adoption of these standards has been seen first in the automotive industry and was supported by IBM through its Manufacturing Integration Framework (MIF) [5].

Meyer [6] presents a complex survey on the intelligent product focusing on the underlying technologies that enable this concept. A classification is introduced that positions an IP along three perspectives: level of intelligence, location of intelligence, and aggregation level of intelligence.

This paper discusses the main classes of shop floor devices relative to distributed intelligence for product-driven automation and traceability and presents in this context an XML approach for: storing the operational information for intelligent products and representing the information flow during manufacturing process, the XSD scheme definition, and the operation dependencies including lead and lag time. We propose and implement a security architecture that uses a public-key infrastructure to generate certificates and propagate trust at runtime (during production execution) for intelligent products on board pallet carriers, communicating with shop floor resources.

2 Classes of SOMA-Enabled Shop Floor Devices

Distributed intelligence and alignment to industry standards are main prerequisites for organizing shop floor activities based on SOA paradigms. The devices used are based on hardware and software standards having a relatively high processing power while adhering to SOA standards; specifically high level communication protocol standards like SOAP are relatively common, as the underlying technology is already available. In contrast, shop floor devices like robots, conveyor belts, CNC machines or feeders are not always equipped with the processing power required for such integration and in this case, the existing hardware needs to be augmented with additional dedicated hardware in order to become a so called SOA-enabled device (Fig. 1).

Recent advances in mobile technology and the emergence of standardized operating systems for mobile devices (such as Android) have made available a large number of low power devices that can be used reliably at shop floor level, allowing implementing the concepts of *distributed intelligence* (DI) in manufacturing and *product-driven automation* to gain traction among practitioners. Depending on the implementation one can identify three main classes of such devices:

Class I: Workstation assisted shop-floor device. This category is represented by the physical device and the associated workstation. In this case the workstation is a standard computer equipped with a dedicated card for connecting to the device. The software is most often proprietary and allows programmatic control of the physical device. The communication protocol between the workstation and the device is proprietary, usually a low level signal based protocol [7–9], see Fig. 2.

For this class of devices, integration in a SOA architecture based on Web services requires creating a software wrapper over the proprietary software APIs provided by the vendor, which would run on the dedicated workstation. Processing power is not a problem in this situation. However, efficiency is an issue as this

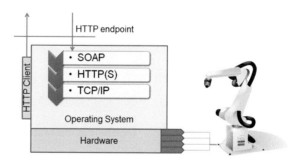

Fig. 1 General architecture of SOA-enabled device for intelligent manufacturing

Fig. 2 Workstation-assisted shop floor device

solution would require usage of a complete workstation equipped with a general purpose operating system. At runtime the utilization of such a workstation is generally less than 5 % of the installed processing power. From a standards perspective this class of devices map to OPC for messaging and ISA-88 for document standards.

Class II: Embedded OS shop-floor device. This class is built up by a hardware environment capable to run an embedded Operating System attached to the shop floor physical device. The requirements for this class are to implement a full HTTP stack, capable to run both a HTTP server for hosting web service endpoints and a HTTP client for calling external web services. The web service in this case is only to expose the existing functionality in SOAP format; it performs data transformation only, Fig. 3.

These devices are generally suitable for mobile shop floor entities, such as pallet carriers transporting products during the manufacturing phases. Efficiency is very high as the embedded OS together with the hardware on which it runs is highly optimized for low power consumption. The standards used by this class of devices are SOAP for messaging protocol and ISA-88 or ISA-95 for document protocols.

Class III: Intelligent shop-floor device. This category of devices is able to run Data and CPU intensive applications in order to implement an intelligent behaviour. For example they are able to run a full Java Virtual Machine on top of the embedded OS and have enough memory and processing power to be able to execute complex algorithms allowing them to make intelligent decisions, such as Genetic Algorithms for scheduling, Neural Networks for decision making and so on (Fig. 4).

Fig. 3 Embedded OS shop floor device

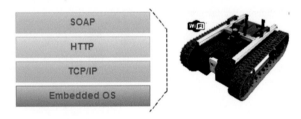

Fig. 4 Intelligent shop floor device

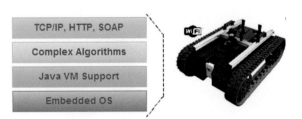

Table 1 Class mapping of manufacturing devices

Device class	Level of intelligence	Location of intelligence	Aggregation level
Class I	All	Remote	Individual
Class II	Problem notification	Remote and local	Individual and container
Class III	All	Local	Individual

In this category one can include mostly the Android equipped devices which can execute complex Java based agents (JADE/WADE). This class of devices are the building blocks for a genuine distributed intelligence architecture, in which complex negotiation logic can be implemented at lower levels in the stack, allowing local decisions in the manufacturing process. These devices can leverage higher layer standards on top of SOAP like ebXML, STEP or OAG BOD.

The manufacturing system architecture can be implemented using any combination of the above devices, as all have in common the generic structure consisting in an informational part and the physical system. From a SOA perspective, the architecture would tend to be a point to point choreography if the lower class devices were used. Once higher class devices introduced, the trend is to use an orchestrated architecture based on BPEL workflows and real time events. The orchestrated architecture offers a high flexibility at integration layer by promoting low coupling between components involved and allows algorithms capable of local decision making. Based on the classification introduced by Meyer [6] in 2009 and discussed in the previous section the capabilities of these device classes are presented in Table 1.

The informational part of the shop floor devices consists at least of structured information regarding both the capabilities of the device for manufacturing resources, and the operations required for intelligent products moving on pallet carriers [10].

3 Data Flows and Structures for Embedded Intelligent Products

The data flow for intelligent products (IP) can be seen as a sequence of steps during the execution of each individual product, as illustrated in Fig. 5.

This process starts when the intelligent product, represented initially only by the pallet carrier equipped with an embedded device, is inserted in the manufacturing line. At this point, the production information is loaded in the memory of the embedded device and initialized. The next step is the data validation activity composed from a XSD schema validation and a logical validation against the operations required for product execution. The logical validation is required in order to detect situations like dead-lock scenarios that might occur. Once the validation is complete each operation from the pre-loaded product recipe is successively executed by shop floor resources. The data structure is updated with the

Fig. 5 IP data flow

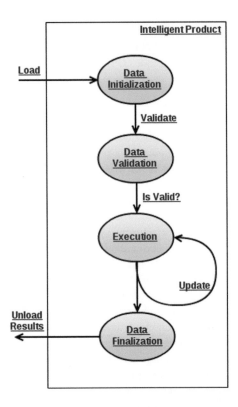

result of each operation execution, until all operations are completed. When the product on pallet exits the manufacturing line, the data finalization phase is executed, where information about each operation execution is consolidated and unloaded from the embedded device on board of the pallet.

The proposed data format is defined using XSD, and contains three sections describing: the product identity, its real time status and the list of operations that need to be executed. The product identity is defined by the following complex type:

```
<complexType name = "ProductIdentity">
<sequence>
<element name = "Batch_ID" type = "string">
</element>
<element name = "Product_ID" type = "string">
</element>
<element name = "RFID_Tag" type = "string">
</element>
<element name = "Product_Type" type = "string">
</element>
```

```
</sequence>
</complexType>
```

The Product_ID together with the RFID_Tag associated to the pallet identify uniquely the product during its execution. The Product_Type represents a pointer to the manufacturing system knowledge base storing the recipe for that specific product.

The real time status of the product is represented by the ProductStatus complex type:

```
<complexType name = "ProductStatus">
<sequence>
<element name = "CriticalPath" type = "string">
</element>
<element name = "TotalEnergyFootprint" type = "float">
</element>
<element name = "Global_EF" type = "float">
</element>
<element name = "Global_LF" type = "float">
</element>
</sequence>
</complexType>
```

The TotalEnergyFootprint is computed in real time during product execution by summing the energy footprint recorded in each operation executed by a resource. The Global_EF and Global_LF (global early finish and late finish estimation) are updated based on the execution path that has been chosen.

The operation list, in other words the execution schedule, is stored in OperationList complex type that is in fact a list of operations:

```
<complexType name = "Operation">
<sequence>
<element name = "ID" type = "string"> </element>
<element name = "Prerequisites" type = "string">
<element name = "Code" type = "string"> </element>
<element name = "Parameters" type = "string">
</element>
<element name = "Resource_ID" type = "string">
</element>
<element name = "Duration" type = "int"> </element>
```

```
<element name = "LeadTime" type = "int"> </element>
<element name = "LagTime" type = "int"> </element>
<element name = "ES" type = "float"> </element>
<element name = "EF" type = "float"> </element>
<element name = "LS" type = "float"> </element>
<element name = "LF" type = "float"> </element>
</element>
<element name = "Energy_Footprint" type = "float">
</element>
<element name = "Quality_Check" type = "string">
<element name = "Operation_Status" type = "string">
</element>
</sequence>
</complexType>
<complexType name = "OperationList">
<sequence>
<element    name    =    "Operation"    type    =    "tns:Operation"
minOccurs = "1"> </element>
</sequence>
</complexType>
```

The Operation complex type is holding the information required for each operation execution. The ID uniquely identifies the operation in the operation list. Prerequisites element is a comma separated list of operations IDs that need to be completed before this operation can start. The Code represents the operation code in the manufacturing system knowledge base and is used in conjunction with the Parameters element. The Resource_ID is the ID of the machine that will execute that specific operation. This value can be known in advance, in case of a pre-computed execution schedule for the whole batch or can be determined at runtime in case of heterarchical mode when there is no predefined execution schedule and job executions are negotiated at runtime.

3.1   Lead and Lag Times for Operations

There are three time-related constraints in the Operation data structure, namely: Duration, Lead Time and Lag Time (see Fig. 6).

- Duration: represents the number of time units required for the operation to be performed by a resource;
- Lag Time: represents the delay imposed between two consecutive operations. For example, if we consider a finish-to-start dependency between Operation1 and

Fig. 6 Lag time and lead
time for operations

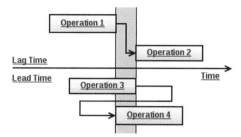

Operation2 with a 5 time units lag time, it would mean that Operation2 cannot
start until at least 5 time units have passed since Operation1 was completed. Lag
time is a positive integer as it adds to the operation's overall duration;

- Lead Time: is an acceleration of the successor operation. In a finish-to-start
  dependency between Operation3 and Operation4 with a 5 time units lead time,
  Operation4 can start up to 5 time units before Operation3 finishes. The lead time
  is expressed as a negative integer because it subtracts from the total duration of
  the operation. These two concepts are illustrated in Fig. 6.

3.2 Early Finish and Late Finish Estimations

Considering the lag time and lead time characteristics for operations together with
the duration and the precedence of operations, one can consider four time estimates
for each operation:

- *ES (early start)*: is the earliest time expressed in time units when the operation
  can start. This is computed based on the prerequisites tree of the operation by
  adding the duration of all operations and the lag time and substantiating the lead
  time. In other words, this represents the optimistic start time for the operation;
- *LS (late start)*: is the latest time expressed in time units when the operation can
  start. Similarly to ES, this is computed form the prerequisite operations, with the
  difference that the lead time is subtracted. This represent the pessimistic start
  time for the operation;
- *EF (early finish)*: is the earliest time when the operation can be finished. It is
  computed as ES + duration of the operation;
- *LF (late finish)*: is the latest time when the operation can be finished. It is
  computed as LS + duration of the operation.

Each product recipe is basically a dependency tree where operations have
generally a pre-imposed precedence. In a hierarchical scheduling mode, where the
execution order of the operations is known in advance, the above four parameters
can be computed directly at the beginning, and during execution are used to validate
that the initial schedule is still being followed by the system.

In a true heterarchical operational mode where the next operation is negotiated at runtime, these parameters are computed at run time, depending on which path in the dependency tree was followed. Computing these parameters at runtime proves to be useful because they enable the estimation of Global_EF and Global_LF of the product, which represents a range for the effective ETA of the product manufacturing process. By considering the Global_EF and Global_LF for all the products in the batch, the Batch_EF and Batch_LF can be computed in the same way.

3.3 Energy Footprint and Quality Check

The Operation complex type has also two attributes that are always added after the operation execution finishes. The Energy_Footprint is added by the resource that just performed the operation, and is a float number representing the energy consumed for the operation. Shop floor resources should be able to report this metric for each predefined operation. The Quality_Check attribute contains the result of the quality control done after each operation. In practice this will be a string status representing either the fact that the check is passed or an error message if a failure was detected.

4 The Security Solution for Active Devices in SOMA

Service orientation of manufacturing control systems (SOMA) has become the standard for manufacturing system design. This approach brings many advantages, proven as best practices by the large scale proliferation especially in the software development industry. In recent years the most important software vendors supporting SOA principles have developed integrated solutions based on SOA technology targeted at manufacturing applications in both research and industry domains.

Multi-agent systems have provided the platform for building distributed SOA architectures in the context of intelligent products equipped with local decision-making capabilities and information storage. This chapter presents a security solution associated with the intelligent shop floor devices in a SOMA oriented architecture and introduces a PKI-based solution designed to secure the information flow against the identified security threats.

4.1 Security Challenges at Shop Floor Level

Considering a shop floor architecture using Wi-Fi communication between intelligent products, resources and production scheduler, there are several challenges regarding the security of the architecture [11–13]. The most relevant for a manufacturing enterprise are:

- *Unauthorized access to information*: an external attacker might get information about customer orders, execution status of products, shop floor resource behaviour and scheduling algorithms. In certain industries gaining access to this information can offer relevant competitive advantages [14, 15];
- *Theft of proprietary information*: product recipes could be reverse engineered based on communication intercepted between shop floor devices, intelligent products and the shop floor scheduler. These recipes usually represent proprietary information [16, 17];
- *Denial of service* (DoS): an external attacker could sabotage the manufacturing system by using denial of service techniques. Depending on the architecture of the manufacturing system communication network and isolation, the vulnerabilities for such an attack can be mitigated [18–20];
- *Impersonation*: the lack of an authentication and authorization mechanism for the shop floor devices raises the challenge of trust when services are invoked. An external attacker might impersonate a shop floor device, in order to gain access to proprietary information or to perform a DoS attack [21, 22].

The main challenge with wireless network security is represented by the simplified network access in contrast with wired networks. In the context of wired networking one must gain access to the facility to physically connect to the internal network. With Wi-Fi networks the attacker only needs to be within the wireless range of the Wi-Fi network in order to be able to attempt an attack. Wi-Fi implementations for sensitive networks usually employ techniques to prevent unauthorized access to the network and at the same time to encrypt the data sent through the network. However, an attacker who has managed to access Wi-Fi network router can initiate a large range of attacks. For example a DNS spoofing attack against any other device in the network is possible by sending a response before the queried DNS server has a chance to reply.

A possible solution to prevent unauthorized access to the Wi-Fi network can be implemented by hiding the access point name by preventing the SSID broadcast. This approach is effective against an uninformed attacker; however it is ineffective as a security method because the SSID is sent in the clear in response to a SSID query. An alternative to this approach, especially for manufacturing systems, where the Wi-Fi devices are known in advance, is to only allow shop floor devices with known MAC addresses to join the network. However, an attacker could eavesdrop the network conversation and join the network by spoofing and replaying an authorized address. Wired Equivalent Privacy (WEP) encryption was designed to protect against casual snooping but it is not considered secure [23, 24]. Tools such as AirSnort or Aircrack-ng can quickly determine WEP encryption keys. As a response to security concerns of WEP, Wi-Fi Protected Access (WPA) was introduced. Even if WEP is more secured then WPA, it still has known vulnerabilities. WPA2 is using Advanced Encryption Standard and eliminates some of the vulnerabilities of WEP. However Wi-Fi Protected Setup which allows initial configuration of the Wi-Fi connection allows WPA and WPA2 security to be broken in several scenarios. Once the network layer is breached, an attacker will have direct access to the higher layer protocols allowing unauthorized access to information, theft of proprietary information, denial of service at the protocol layer and impersonation [25, 26]. The higher layer protocols can be secured using secure sockets layer (SSL) that provides encryption, authentication and authorization of the actors involved using a *public-key infrastructure* (PKI) for certificate management.

The problems related to security at the shop floor control layer and middleware were mentioned by previous research [27] together with approaches to tackle them in general terms [28, 29]. Some of the most relevant are further mentioned.

SOCRADES, a Web Service based Shop Floor Integration Infrastructure proposed by SAP Research [30] in 2008, defines two specific security requirements for shop floor devices: a) to be able to authenticate themselves to external services and b) to authenticate/control access to services they offer. The solution proposed is based on role-based access control principle of devices communication to middleware and back end services. Additionally, message integrity and confidentiality is provided by the WS-Security standard.

iShopFloor is an intelligent shop floor based on the Internet, web, and agent technologies. It focuses on implementing distributed intelligence in the manufacturing shop floor [31]. The iShopFloor research outlines some security challenges related to usage of secure socket layer (SSL) in an open environment. The main problems are caused by SSL termination in the frontend servers and secure communication by firewalls. Another research [32] describes a CORBA-based integration framework for distributed shop floor control. In this implementation the security is provided by the object broker as a platform service, including authentication and authorization in a centralized fashion.

The PKI solution presented in this paper has several advantages compared to the ones presented above:

- It directly addresses the shop floor Wi-Fi enabled devices (intelligent products), rather than concerning about intra middleware software communication. The middleware is normally located in a physical secure location and uses wired communication within the DMZ. This makes it almost impossible for an attacker to intercept communication without physically breaking in.
- It provides a real time mechanism to allow certificate generation and revocation. This mechanism can be integrated at CA layer with RBAC solutions if required.
- It is a light implementation, suited for mobile devices (embedded device mounted on the product pallet).

4.2 SSL and PKI Implementation for Manufacturing Shop Floor Devices

The Secure Socket Layer protocol, first introduced by Netscape, was used initially to ensure secure communication between web servers and browsers communicating over HTTP(s). However, the protocol itself is at socket layer, so the applications are not limited to HTTP.

The protocol proposed in the paper uses a PKI infrastructure, consisting in a Certificate Authority (CA) that generates public-private key pairs to identify the server, the client or both. The SSL handshake, in case of mutual authentication, consists in the following main steps:

1. The client connects to a secure server socket.
2. The server sends its public key with its certificate.
3. The client checks if the certificate was issued by a trusted CA, that the certificate is still valid and that the certificate is matching the DNS hostname of the server.
4. The client passes its public key with its certificate to the server. The server verifies if the client certificate was issued by a trusted CA, similarly with step 4.
5. The server responds with a negotiation request on the encryption cypher used.
6. The client then uses the public key to encrypt a random symmetric encryption key and sends it to the server.
7. The server decrypts the symmetric encryption key using its private key and uses the symmetric key to decrypt data.
8. The server sends back the data encrypted with the symmetric key.
9. Communication continues encrypted until the socket is closed.

In order to use SSL with mutual certificate authentication, a PKI is required. The PKI is considered as a software platform that basically generates public/private key pairs and certificates and associates them with identities by means of a certificate authority (CA).

The proposed security architecture involves using a public-key infrastructure PKI to generate certificates and propagate trust during manufacturing execution (runtime) for intelligent products communicating with shop floor resources. The sequence diagram for trust propagation is presented in Fig. 7.

The process is divided in two separate phases. The pre-configuration phase consists in generating the certificates for the shop floor resources that are static in nature, or in other words, are present during the entire manufacturing process. Examples of such resources are robots equipped with Web Service capabilities, the shop floor scheduler or the PLC for conveyor service. These certificates are valid for a long period of time, as they belong to these static resources. During this phase, the trust is configured for the shop floor static devices, by importing the CA public key in the local trust-store.

The run-time phase refers to the products equipped with local web service capabilities. When the intelligent product enters the production line the embedded device on the product pallet is initialized.

The main steps are creation of certificate request by the IP, generation of the certificate by the CA, creation of the certificate store on the intelligent product, product execution and certificate revocation when the product is completed. The following sections exemplify these steps using OpenSSL as PKI infrastructure and JADE agents for intelligent product implementation.

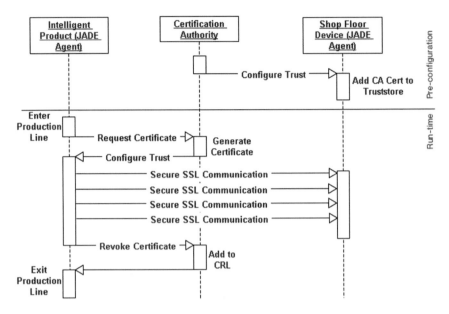

Fig. 7 Sequence diagram for trust propagation

4.3 Certificate Generation at Runtime Using OpenSSL and JADE Agents

The PKI implementation proposed for securing service oriented manufacturing systems is based on OpenSSL [33]. OpenSSL is an Open Source toolkit implementing the Secure Sockets Layer (SSL v2/v3) and Transport Layer Security (TLS v1) protocols as well as a full-strength general purpose cryptography library. The internal JADE agent architecture when integrated with OpenSSL is illustrated in Fig. 8.

The SOA enabled device must be capable of both inbound and outbound SOAP capabilities. For run time encryption the Java SSL implementation is used. However, for initial configuration and for the CA agent, the OpenSSL native libraries are used. The native library is accessed using a JNI wrapper library.

Step 1 Certificate request: consists in **calling X509_REQ_NEW function to create a X.509 certificate request.** An example data structure passed to the method is:

```
struct entry entries[ENTRIES] =
{
{"countryName", "RO" },
```

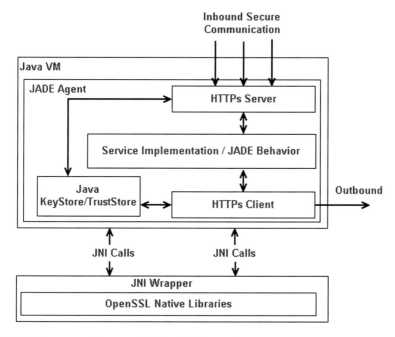

Fig. 8 JADE agent architecture with OpenSSL integration

{"stateOrProvinceName", "RO" },
{"localityName", "Bucharest" },
{"organizationName", "University Politehnica Bucharest" },
{"organizationalUnitName", "CIMR" },
{"commonName", <DNS name of the device based on RFID tag>},
};

The commonName must match to the DNS name of the embedded device, in order to pass host name verification during SSL handshake. To uniquely identify the product, the host name is constructed using the individual RFID tag from the product pallet. Once the certificate request is completed, it is sent to the CA agent.

Step 2 Certificate signing: CA agent receives the certificate request and generates the X509 certificate that will be used by the IP. The certificate is installed in the local key-store on the product and the CA agent public key is added to the trust-store.

Step 3 Execution: at this stage the agent is able to securely communicate and authen-ticate with SSL mode 3 (which includes mutual authentication) with other shop floor devices based on CA trust established. The sequence of messages exchanged between the client and the server (shop floor actors) during SSL handshake at runtime, are:

Client 1: Client sends a CLIENT_HELLO command to the server, including:

- The SSL and TLS version supported by the client
- A set of ciphers supported by the client in order of preference
- A set of data compression methods supported by the client
- The session ID, which is 0 in case of a new SSL session
- A segment of random data generated by the client for key generation

Server 1: Server sends back a SERVER_HELLO command to the client, which includes:

- The SSL or TLS version that will be used for the SSL session
- The cipher selected by the server that will be used for the SSL session
- The data compression method selected by the server
- The generated session ID for the SSL session
- A segment of random data generated by the server for key generation

Server 2: Server sends the CERTIFICATE command:

- This command includes the server certificate
- The client will validate the certificate against the truststore, check the hostname against the certificate CN and check the Certificate Revocation List from the CA

Server 3: Server sends the CERTIFICATE_REQUEST command to request the client certificate. The command contains the certificate authorities (CAs) that the server trusts, allowing the client to send the corresponding certificate.

Client 2: The client sends the CERTIFICATE command, sending its certificate to the server. The server will validate the client certificate against the truststore, check the hostname against the client certificate CN and the Certificate Revocation List from the CA.

Client 3: The client sends the CERTIFICATE_VERIFY command to the server, which contains a digest of the SSL handshake messages signed using the client private key. The server also calculates the digest using the client's public key from the client certificate. The two are compared by the server and if they match the client is verified.

Server 4: Server sends the SERVER_DONE command, indicating that the server has completed the SSL handshake.

Client 4: The client sends the CLIENT_KEY_EXCHANGE command

This command contains the premaster secret that was created by the client and was then encrypted using the server public key. Both the client and the server generate the symmetric encryption keys on their own using the premaster secret and the random data generated from the SERVER_HELLO and CLIENT_HELLO commands.

Client 5: The client sends the CHANGE_CIPHER_SPEC command, indicating that the contents of subsequent SSL record data sent by the client during the SSL session will be encrypted with the selected cipher.

Client 6: The client sends the FINISHED command, including a digest of all the SSL handshake commands that have flowed between the client and server. This command is used to confirm that none of the commands exchanged so far were tampered with

Server 5: The server sends the CHANGE_CIPHER_SPEC command, indicating the cipher used to encrypt following SSL messages.

Server 6: The server sends the FINISHED command, including a digest of all the SSL handshake commands that have flowed between the server and the client. This command is used to confirm that none of the commands exchanged so far were tampered with. From this stage on, the messages exchanged are encrypted and secure.

Step 4 Revocation: is the last stage after the product execution is completed. As the product will no longer require communication with other shop floor devices, the certificate must be revoked. This is accomplished by sending a certificate revocation request to the CA agent. The CA agent publishes the certificate revocation in the CRL list, so that all future SSL handshakes that use this certificate will be prevented

The experimental evaluation of the security provided is presented in Sect. 5.

5 Experimental Results: PKI and Network Trace Evaluation

A simulation environment employing JADE agents for IP and shop floor resources was created to evaluate the PKI solution proposed and analyse network trace (Fig. 9).

The simulation environment consists in two agent instances communicating over the WEP encrypted Wi-Fi network. The interaction between two agents is considered using the described PKI setup, assuming an attacker that was already able to break the WEP encryption and gain access to the network. The attacker uses AirPcapNG to intercept the Wi-Fi traffic and analyse it using Wireshark tool. Figure 9 presents the network traffic obtained by the attacker in this scenario. The network analysis shows the certificate exchange with both server and client authentication at SSL/TLS layer. The last packet in the above trace shows the beginning of the encrypted application data conversation between two agents.

Looking closer at SERVER_HELLO the server certificate together with the CA certificate are being sent to the client, as highlighted in Fig. 10.

By implementing the PKI infrastructure for SSL communication presented in this paper, some important security challenges are mitigated. The asymmetric encryption in SSL assures that un-authorized access to information is prevented even if the Wi-Fi packets are captured by a potential attacker. The encryption also prevents access to proprietary information that might be stored and communicated between intelligent products, shop floor scheduler and resources. The possibility of Denial of Service attacks is not completely eliminated by this approach; however

Fig. 9 Wireshark view of the encrypted traffic (Wireshark analysis)

```
⊞ Handshake Protocol: Server Hello
⊟ Handshake Protocol: Certificate
     Handshake Type: Certificate (11)
     Length: 1703
     Certificates Length: 1700
   ⊟ Certificates (1700 bytes)
       Certificate Length: 854
     ⊞ Certificate (id-at-commonName=HMES_EA32454F1A2DDA11 id-at-organizationalUnitName=CIMR,id-at-organiz
       Certificate Length: 840
     ⊞ Certificate (id-at-commonName=HMAS_CA id-at-organizationalUnitName=CIMR,id-at-organizationName=PUB,
```

Fig. 10 Detail of server hello (Wireshark analysis)

they are limited to the network layer due to SSL implementation. Impersonation is also prevented by implementing mutual SSL authentication during initial handshake. One limitation at this point is represented by the requirement for Step 1 and 2 of the initialization process to be performed over wired network to prevent certificate spoofing during private key transmission.

6 Conclusions and Perspective of Further Research

The paper proposes a classification of intelligent products from SOA integration point of view and introduces a formalized data structure for intelligent products in the form of a XSD schema for XML representation. The data flow during manufacturing is discussed in the context of lead- and lag time between operations in the product recipe in order to enable ETA estimation of the product and the product batch; experiments focus on representing operation dependencies and lag time in XML format.

A PKI solution with SSL authentication and encryption for intelligent products travelling on pallets in the shop floor has been designed. From an implementation perspective SOA alignment at shop floor level involves TCP/IP-based communication over Wi-Fi supporting higher level protocols like SOAP for Web service based integration. Our contribution in the security area of SOMA implementation presents security challenges associated with this approach, such as information flow protection as well as authentication and authorization of the actors involved. A public-key infrastructure (PKI) was proposed to assure both SSL encryption and authentication services at the shop floor layer that protects the manufacturing system against possible external attacks. The solution consists in a shop floor certification authority (CA) agent that dynamically generates certificates for intelligent products travelling on the production line. A certificate revocation list (CRL) is used to revoke the individual certificates once the product is completed and exits the production line. The prototype implementation of this PKI solution, based on OpenSSL encryption libraries, is presented in the context of a JADE MAS shop floor design.

Future work is aiming at evaluating other forms of attacks that this PKI architecture might be exposed at in an industrial environment, and evaluating the overall performance overhead introduced by the encryption and decryption of the information flow across shop floor devices.

References

1. Wong, C. Y., McFarlane, D., Zaharudin, A., Agarwal, V.: The intelligent product driven supply chain. In: IEEE International Conference on IEEE Systems, Man and Cybernetics, vols. 4, 6 (2002)
2. McFarlane, D., Sarma, S., Chirn, J.L., Ashton, K.: The intelligent product in manufacturing control and management. In: 15th Triennial World Congress, Barcelona (2002
3. Morariu, C., Borangiu, T.: Manufacturing integration framework: a SOA perspective on manufacturing. In: Proceeding of Information Control Problems in Manufacturing (INCOM'12), IFAC Papers OnLine, vol. 14, no. 1, 31–38, (2012) (Elsevier)
4. Främling, K., Harrison, M., Brusey, J., Petrow, J.: Requirements on unique identifiers for managing product lifecycle information: comparison of alternative approaches. Int. J. Comput. Integr. Manuf. **20**(7), 715–726 (2007)

5. Zhang, L.-J., Nianjun, Z., Chee, Y.-M., Jalaldeen, A., Ponnalagu, K., Arsanjani, A., Bernardini, F.: SOMA-ME: a platform for the model-driven design of SOA solutions. IBM Syst. J **47**(3), 397–413 (2008)
6. Meyer, G.G., Främling, K., Holmström, J.: Intelligent products: a survey. Comput. Ind. **60**(3), 137–148 (2009). Elsevier
7. Kiritsis, D., Bufardi, A., Xirouchakis, P.: Research issues on product lifecycle management and information tracking using smart embedded systems. Adv. Eng. Inform. **17**(3), 189–202 (2003)
8. Jun, H.-B., Shin, J.-H., Kim, Y.-S., Kiritsis, D., Xirouchakis, P.: A framework for RFID applications in product lifecycle management. Int. J. Comput. Integr. Manuf. **22**(7), 595–615 (2009)
9. Leitão, P.: Agent-based distributed manufacturing control: a state-of-the-art survey. Eng. Appl. Artif. Intell. **22**(7), 979–991 (2009)
10. Ventä, O.: Intelligent products and systems: Technology theme-final report. VTT Technical Research Centre of Finland (2007)
11. Potter, B.: Wireless hotspots: Petri dish of wireless security. Commun. ACM **49**(6), 50–56 (2006)
12. Berghel, H., Uecker, J.: Wireless infidelity II: air jacking. Commun. ACM **47**(12), 15–20 (2004)
13. Boncella, R.J.: Wireless security: an overview, Commun. Assoc. Info. Syst. **9**(15), 269–282 (2002)
14. Von Solms, B., Marais, E.: From secure wired networks to secure wireless networks–what are the extra risks? Computers and Security **23**(8), 633–637 (2004)
15. Kankanhalli, A., Teo, H.-H., Tan, B., Wei, K.-K.: An integrative study of in-formation systems security effectiveness. Int. J. Info. Manage. **23**(2), 139–154 (2003)
16. Whitman, M.E.: Enemy at the gate: threats to information security. Commun. ACM **46**(8), 91–95 (2003)
17. Mercuri, R.: Analyzing security costs. Commun. ACM **46**(6), 15–18 (2003)
18. Wang, Y., Chuang, L., Quan-Lin, L., Fang, Y.: A queuing analysis for the denial of service (DoS) attacks in computer networks. Comput. Netw. **51**(12), 3564–3573 (2007)
19. Warren, M., Hutchinson, W.: Cyber-attacks against supply chain management systems. Int. J. Phys. Distrib. Logistics Manage. **30**(7/8), 710–716 (2000)
20. Reaves, B. Morris, T.: Discovery, infiltration, and denial of service in a process control system wireless network. In: eCrime Res. Summit, IEEE 1–9 (2009)
21. Peine, H.: Security concepts and implementation in the Ara mobile agent system. In: 7th IEEE International Workshops on IEEE Enabling technologies: infrastructure for collaborative enterprises (WET ICE'98 Proceedings), pp. 236–242 (1998)
22. Nilsson, D., Larson, U., Jonsson, E.: Creating a secure infrastructure for wireless diagnostics and software updates in vehicles. In: Computer Safety, Reliability, and Security, pp. 207–220. Springer (2008)
23. Boland, H., Mousavi, H.: Security issues of the IEEE 802.11 b wireless LAN. In: Canadian Conference on IEEE, El. and Computer Engineering, **1**, 333–336 (2004)
24. Reddy, S., Vinjosh, K., Sai R., Rijutha, K., Ali, S.M., Reddy, C.P.: Wireless hacking-a WiFi hack by cracking WEP. In: 2nd International Conference on IEEE, Education Technology and Computer (ICETC), **1**, V1–189 (2010)
25. Berghel, H., Uecker, J.: WiFi attack vectors. Comm. ACM **48**(8), 21–28 (2005)
26. Aime, M.D., Calandriello, G., Lioy, A.: Dependability in wireless networks: can we rely on WiFi?, Security & Privacy, IEEE **5**(1), 23–29 (2007)
27. Wang, L., Orban, P., Cunningham, A., Lang, S.: Remote real-time CNC machining for web-based manufacturing. Robot. CIM **20**(6), 563–571 (2004)
28. Wang, L., Shen, W.,Lang, S.: Wise-ShopFloor: a web-based and sensor-driven shop floor environment. In: The 7th International Conference on IEEE Computer Supported Cooperative Work in Design, pp. 413–418 (2002)

29. Sauter, T.: The continuing evolution of integration in manufacturing automation, Ind. Electron. Mag. IEEE **1**(1), 10–19 (2007)
30. De Souza, L. et al.: Socrades: A web service based shop floor integration infrastructure. The Internet of Things, pp. 50–67 (2008)
31. Shen, W., Lang, S.Y.T., Wang, L.: iShopFloor: an Internet-enabled agent-based intelligent shop floor, Systems, Man, and Cybernetics, Part C: Applications and Reviews, IEEE Trans. on **35**(3), 371–381 (2005)
32. Shin, J., et al.: CORBA-based integration framework for distributed shop floor control. Comput. Ind. Eng. **45**(3), 457–474 (2003)
33. OpenSSL.: http://openssl.org/. Accessed June 2015-04-23

Part VI
Cloud and Computing-Oriented Manufacturing

Technological Theory of Cloud Manufacturing

Sylvain Kubler, Jan Holmström, Kary Främling and Petra Turkama

Abstract Over the past decade, a flourishing number of concepts and architectural shifts appeared such as the Internet of Things, Industry 4.0, Big Data, 3D printing, etc. Such concepts are reshaping traditional manufacturing models, which become increasingly network-, service- and intelligent manufacturing-oriented. It sometimes becomes difficult to have a clear vision of how all those concepts are interwoven and what benefits they bring to the global picture (either from a service or business perspective). This paper traces the evolution of the manufacturing paradigms, highlighting the recent shift towards Cloud Manufacturing (CMfg), along with a taxonomy of the technological concepts and technologies underlying CMfg.

Keywords Cloud manufacturing · Internet of things · Direct digital manufacturing

S. Kubler (✉)
Interdisciplinary Centre for Security Reliability and Trust,
University of Luxembourg, 2721 Luxembourg, Luxembourg
e-mail: sylvain.kubler@uni.lu

J. Holmström
Department of Civil and Structural Engineering, School of Engineering,
Aalto University, P.O. Box 11000, 00076 Aalto, Finland
e-mail: jan.holmstrom@aalto.fi

K. Främling
Department of Computer Science, School of Science, Aalto University,
P.O. Box 15500, 00076 Aalto, Finland
e-mail: kary.framling@aalto.fi

P. Turkama
Center for Knowledge and Innovation Research, School of Business,
Aalto University, P.O. Box 15500, 00076 Aalto, Finland
e-mail: petra.turkama@aalto.fi

© Springer International Publishing Switzerland 2016
T. Borangiu et al. (eds.), *Service Orientation in Holonic and Multi-Agent Manufacturing*, Studies in Computational Intelligence 640,
DOI 10.1007/978-3-319-30337-6_24

267

1   Introduction

Manufacturing paradigms evolved over time, driven by societal trends, new ICT (information and communication technology) technologies, and new theories. The manufacturing processes of the future need to be highly flexible and dynamic in order to map the customer demands, e.g. in large series production or mass customization. Manufacturing companies are not only part of sequential, long-term supply chains, but also of extensive networks that require agile collaboration between partners. Companies involved in such networks must be able to design, configure, enact, and monitor a large number of processes and products, each representing a different order and supply chain instance. One way of achieving this goal is to port essential concepts from the field of Cloud Computing to Manufacturing, such as the commonly applied SPI model: SaaS (Software-as-a-Service), PaaS (Platform-as-a-Service), IaaS (Infrastructure-as-a-Service) [1]. In the literature, this concept is referred to as "Cloud manufacturing" (CMfg), which has the potential to move from production-oriented manufacturing processes to customer- and service-oriented manufacturing process networks [2], e.g. by modelling single manufacturing assets as services in a similar vein as SaaS or PaaS solutions.

While organizations will be looking to make use of CMfg for creating radical change in manufacturing practices, this will not be an easy transition for many. There will be architectural issues as well as structural considerations to overcome. The main reason for this is that CMfg derives not only from cloud computing, but also from related concepts and technologies such as the Internet of Things—IoT (core enabling technology for goods tracking and product-centric control) [3, 4], 3D modelling and printing (core enabling technology for digital manufacturing) [5, 6], and so on. Furthermore, some of those concepts/technologies have not yet reached full maturity such as the IoT, whose number of connected devices should pass from 9.1 billion (2013) to 28.1 billion (2020) according to IDC forecasts). Similarly, while 3D modelling is now conventional even for small companies, 3D printing is still in the peak of inflated expectation phase in the Gartner Hype Cycle, which may be (potentially) followed by a drop into the trough of disillusionment [7]. Within this context, the success of CMfg is partly dependent upon the evolution of all these concepts, although it is often difficult to understand how they are interwoven and how important one is to the other. The present paper helps to better understand such interwoven relationships, the current trends and challenges (e.g., shift from closed-industry solutions to open infrastructures and marketplaces).

To this end, Sect. 2 shows the evolution of the manufacturing paradigms through the ages. Section 3 introduces a CMfg taxonomy, whose key challenges and opportunities of the underlying concepts are discussed; the conclusions follow.

2 Manufacturing Paradigms Through the Ages

Over the last two centuries, manufacturing industry has evolved through several paradigms from Craft Production to CMfg [8, 9]. Craft Production, as the first paradigm, responded to a specific customer order based on a model allowing high product variety and flexibility, where highly skilled craftsmen treated each product as unique. However, such a model was time- and money-consuming—*as depicted in 1*. The history of production systems truly began with the introduction of standardized parts for arms, also known as the "American System" (see Fig. 1).

Following the American System model, Mass Production enabled the making of products at lower cost through large-scale manufacturing. On the bad side, the possible variety of products was very limited since the model is based on resources performing the same task again and again, leading to significant improvement of speed and reduction of assembly costs (*cf.* 1). Symbols for mass production were Henry Ford's moving assembly line and his statement: *"Any customer can have a car painted any color that he wants so long as it is black"*.

Lean Manufacturing emerged after World War II as a necessity due to the limited resources in Japan. The Lean Manufacturing paradigm is a multi-dimensional approach that encompasses a wide variety of management practices, including just-in-time, quality systems, work teams, cellular manufacturing, etc., in an integrated system [10] that eliminates "waste" on all levels. It is worth noting that the lean management philosophy is still an important part of all modern production systems.

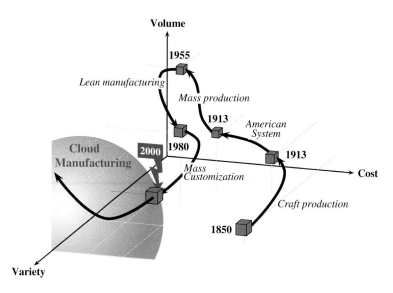

Fig. 1 Volume-variety-cost relationship in manufacturing paradigms

The fourth paradigm, Mass Customization, came up in the late 1980s when the customer demand for product variety increased. The underlying model combines business practices from Mass Production and Craft Production, moving towards a customer-centric model. This model requires the mastery of a number of technologies and theories to make manufacturing systems intelligent, faster, more flexible, and interoperable. Within this context, a significant body of research emerged, particularly with the IMS (Intelligent Manufacturing System) community with worldwide membership, which is an industry-led, global, collaborative research and development program established to develop the next generation of manufacturing and processing technologies. The IMS philosophy adopts heterarchical and collaborative control as its information system architecture [11–13]. The behaviour of the entire manufacturing system therefore becomes collaborative, determined by many interacting subsystems that may have their own independent interests, values, and modes of operation.

It is clear from Fig. 1 that the manufacturing paradigms succeeded one another, always seeking for smaller volumes and costs, while rising the product variety. The fifth and recent paradigm, CMfg, moves this vision a step further since it provides service-oriented networked product development models in which service consumers are enabled to configure, select, and use customized product realization resources and services, ranging from computer-aided engineering software to reconfigurable manufacturing systems [14, 15]. Several applications relying on Cloud infrastructure have been reported in recent years, e.g. used for hosting and exposing services related to manufacturing such as machine availability monitoring, collaborative and adaptive process planning, online tool-path programming based on real-time machine monitoring, collaborative design, etc. [16, 17]. Similarly in the European sphere, this technology has recently attracted a lot of attention, e.g. with the Future Internet Public Private Partnership (FI-PPP),[1] OpenStack, OpenIoT,[2] or Open Platform 3.0 communities.[3]

The next section helps to understand what concepts and technologies are underlying CMfg, how they are interwoven together, how important one is to the other, and what challenges remain ahead.

3 Cloud Manufacturing Taxonomy

The Industrial Internet, Industry 4.0, CMfg, or still Software Defined Manufacturing (SDM) are terms referring to the new phenomenon (or next wave) of innovation impacting the way the world connects and optimizes machines, as well as information systems in the manufacturing industry. In CMfg applications,

---

[1]http://www.fi-ppp.eu.

[2]https://github.com/OpenIotOrg/openiot/wiki/OpenIoT-Architecture.

[3]http://www.opengroup.org/subjectareas/platform3.0.

various manufacturing resources and abilities can be intelligently sensed and connected into a wider Internet, and automatically managed and controlled using both (either) IoT and (or) Cloud solutions, as emphasized in the taxonomy given in Fig. 2. In this taxonomy, one can see that the so-called IoT is a core enabler, if not the cornerstone, for product-centric control and increasing servitization (i.e., making explicit the role of the product as the coordinating entity in the delivery of customized products and services) [18]. Product-centric control methods are, in turn, required and of the utmost importance for developing fast and cost effective Direct Digital Manufacturing (DDM) solutions [6], also known as `Rapid Manufacturing'. One example of how CMfg platforms combine all those concepts might be the following:

> A tractor (or backend system) detects – based on sensor data fusion – that the pump is defective. The after-sales service system is immediately notified and turns to the services of the cloud manufacturing community to (i) access product-related data and models (e.g., CAD models) and then (ii) identify an optimal manufacturer for the broken pump parts. The digital model is sent to the community member who can produce the custom part via 3D printing. The closest (or cheapest) 3D printer service provider(s) can be discovered (e.g., via IoT discovery mechanisms), so that the pump part can be produced to order and shipped to the farmer.

Sections 3.1–3.4 discusses in greater detail all the taxonomy concepts and interdependencies, along with challenges that still need to be addressed.

3.1 Cloud Computing

Cloud computing has revolutionized the way computing infrastructure is abstracted and used [1]. The benefits of Cloud for manufacturing enterprises are numerous; Cloud as a procurement model delivers undisputed cost efficiencies and flexibility, while increasing reliability, elasticity, usability, scalability and disaster recovery. A key difference between Cloud computing and CMfg is that resources involved in

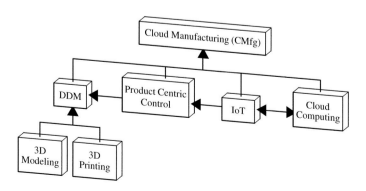

Fig. 2 CMfg taxonomy: underlying concepts and technologies

cloud computing are primarily computational (e.g., server, storage, network, software), while in CMfg, all manufacturing resources and abilities involved in the whole life cycle of manufacturing are aimed to be provided for the user in different service models [2]. The manufacturing resources and abilities are virtualized and encapsulated into different manufacturing cloud services, where different product stakeholders can search and invoke the qualified services according to their needs, and assemble them to be a virtual manufacturing environment or solution to complete their manufacturing task [15].

As an end consumer looking at the cloud space, there are two major types of clouds to choose from: open source clouds (e.g., Citrix, OpenIoT) and closed clouds (e.g., Amazon, Azure, Google). One of the key challenges, especially from the EU perspective, is to foster cloud manufacturing based on existing open standards and components to facilitate an as-vendor-independent-as-possible Cloud engineering workflows platform should lead to radical transformations in business dynamics in the industry (e.g., for new open standard-based value creation) [19, 20]. This implies creating cloud manufacturing ecosystem(s) built on open IoT messaging standards having the capabilities to achieve "Systems-of-Systems" integration, as will be discussed in the next section.

3.2 Internet of Things (IoT)

The growth of the IoT creates a widespread connection of "Things", which can lead to large amounts of data to be stored, processed and accessed. Cloud computing is one alternative for handling those large amounts of data. To a certain extent, the cloud effectively serves as the brain to improve decision-making and optimization for IoT-connected objects and interactions [21], although some of those decisions can be made locally (e.g., by the product itself) [12, 13]. However, as stated previously, new challenges arise when IoT meets Cloud; e.g. creating novel network architectures that seamlessly integrate smart connected objects, as well as distinct cloud service providers (as illustrated with the dashed arrows in Fig. 3). IoT standards e.g. for RESTful APIs and associated data will be key to be able to import/export product-related data and models inside CMfg ecosystems [22].

Several research initiatives have addressed this vision such as—*in the EU sphere* —the IERC or FI-PPP clusters (see e.g. FI-WARE, OpenIoT), or still the Open Platform 3.0 (initiative of The Open Group). In this respect, our research claims that the recent IoT standards published by The Open Group, notably O-MI and O-DF [3], have the potential to fulfill the "Systems-of-Systems" vision discussed above. O-MI provides a generic Open API for any RESTful IoT information system, and O-DF is a generic content description model for Objects in the IoT, which can be extended with more specific vocabularies (e.g., using or extend domain-specific

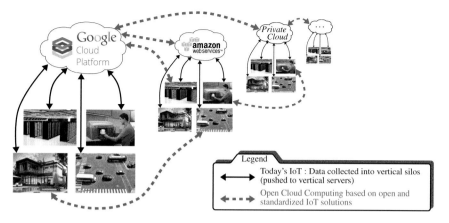

Fig. 3 Challenge of creating CMfg ecosystem based on open IoT standards

ontology vocabularies). Both standards are about to be used as foundation of the upcoming H2020 project bIoTope (Building an IoT OPen innovation Ecosystem for connected smart objects), where proofs of concept and value of open CMfg ecosystems will likely be developed. Furthermore, O-MI and O-DF specifications were identified from several real-life industrial applications of the PROMISE EU project (including manufacturing scenarios) [23], thus making it suitable for effective Product Centric Control, as will be discussed in the next section.

3.3   Product Centric Control

In a true IoT, each intelligent product and equipment is uniquely identifiable [24], making it possible to link control instructions with a given product-instance. The basic principle is that the product itself, while it is in the process of being produced and delivered, directly requests processing, assembly and materials handling from available providers, therefore simplifying materials handling and control, customization, and information sharing in the supply chain. This concept is referred to as "Product Centric Control" [25], which is required and of the utmost importance from a CMfg perspective since it allows for developing fast and cost effective DDM solutions, as will be discussed in the next section. Indeed, operations and decision making processes that are triggered and controlled by the product itself result in higher quality and efficiency than standard operations and external control. The generative mechanism is somehow the ability of the product to (i) monitor its own status; (ii) notify the user when something goes wrong (e.g., the defective pump); (iii) help the user to find and access the necessary product-related models and information from the manufacturer community involved in the CMfg ecosystem; and (iv) ease the synchronization of product-related data and models that might be generated in distinct organizations, throughout the product lifecycle [12, 26].

3.4 Direct Digital Manufacturing—DDM

Recently, the range of DDM[4] technologies has increased significantly with the advancement of 3D printing [6], opening up a novel range of applications considered impossible, infeasible or uneconomic in the past. DDM technologies are technologies that include both novel 3D printing and 3D modelling (as emphasized in Fig. 2), i.e. the more conventional numerical controlled machines. The need for tooling and setup is reduced by producing parts directly based on a digital model. The implication of the development of DDM technologies is that, in an increasing number of situations, it is possible to produce parts directly to demand, without tooling, setup and consideration of economies of scale [27]. Time-to-market, freedom of design, freedom to redesign and flexible manufacturing plans are only the beginning. These advantages represent just the tip of the iceberg since DDM is a relatively new manufacturing practice.

Given this, CMfg is clearly an applicable business model for 3D-printing. Because additive manufacturing is a digital technique, it is possible to manufacture products close to the location where they will be used, thus reducing transportation ($Co2$ emissions), large storage areas, while enabling a wide range of customers, suppliers and manufacturers to take part to the development of new products and services based on an open and standardized CMfg platform.

4 Conclusion

In industry, cloud manufacturing (CMfg) platforms are rarely applied today because of considerable concerns about security and ROI (due mainly to considerable efforts to implement interoperability). Furthermore, the maturity of the platforms is often limited to a prototype status nowadays. However, there are some industry settings, from which interest in such a concept is stated such as associations of SMEs who intend to jointly provide customisable products, or industry clusters who would like to make their members' abilities easily available (searchable and usable) for other members.

Within this context, the emergence of the Internet of Things, Cloud computing, 3D printing, product-centric-control techniques, etc., mark a new turning point for CMfg—manufacturing resources and organization assets become easier to be remotely tracked, monitored, accessed, booked and used (e.g., for production), when and as needed. However, all those concepts make it difficult to understand how they are interwoven and what benefits they bring to the global picture (either from a service or business perspective). This paper contributes to the discussion about this global picture with the introduction of a CMfg taxonomy, while discussing current trends and challenges that still face CMfg (e.g., shift from

---

[4]DDM is the usage of additive manufacturing for production of end-use components.

closed-industry solutions to open infrastructures and marketplaces). In this regard, this paper claims that the vision of "Systems-of-Systems" built on open standards (e.g., open IoT standards as O-MI/O-DF) will be key in the future to develop more advanced open- and customer-oriented CMfg models, which will result in innovative business transformation services.

Acknowledgements The research leading to this publication is supported by the National Research Fund Luxembourg (grant 9095399) as well as Academy of Finland (Grant 275839).

References

1. Mell, P., Grance, T.: The nist definition of cloud computing. Technical report, National Institute of Standards and Technology (2011)
2. Li, B.H., Zhang, L., Wang, S.L.: Cloud manufacturing: a new service-oriented networked manufacturing model. Comput. Integr. Manuf. Syst. **16**(1), 1–16 (2010)
3. Främling, K., Kubler, S., Buda, A.: Universal messaging standards for the iot from a lifecycle management perspective. IEEE Internet Things J. **1**(4), 319–327 (2014)
4. Cai, H., Xu, L.D., Xu, B., Xie, C., Qin, S., Jiang, L.: Iot-based configurable information service platform for product lifecycle management. IEEE Trans. Industr. Inf. **10**(2), 1558–1567 (2014)
5. Berman, B.: 3-d printing: the new industrial revolution. Bus. Horiz. **55**(2), 155–162 (2012)
6. Khajavi, S.H., Partanen, J., Holmström, J.: Additive manufacturing in the spare parts supply chain. Comput. Ind. **65**(1), 50–63 (2014)
7. Kietzmann, J., Pitt, L., Berthon, P.: Disruptions, decisions, and destinations: Enter the age of 3-d printing and additive manufacturing. Bus. Horiz. **58**(2), 209–215 (2015)
8. Clarke, C.: Automotive Production Systems and Standardisation: From Ford to the Case of Mercedes-Benz. Springer Science & Business Media (2005)
9. Herrmann, C., Schmidt, C., Kurle, D., Blume, S., Thiede, S.: Sustainability in manufacturing and factories of the future. Int. J. Precis. Eng. Manuf. Green Technol. **1**(4), 283–292 (2014)
10. Shah, R., Ward, P.T.: Lean manufacturing: context, practice bundles, and performance. J. Oper. Manage. **21**(2), 129–149 (2003)
11. Van Brussel, H., Wyns, J., Valckenaers, P., Bongaerts, L., Peeters, P.: Reference architecture for holonic manufacturing systems: PROSA. Comput. Ind. **37**(3), 255–274 (1998)
12. Meyer, G., Främling, K., Holmström, J.: Intelligent products: a survey. Comput. Ind. **60**(3), 137–148 (2009)
13. McFarlane, D., Giannikas, V., Wong, A.C.Y., Harrison, M.: Product intelligence in industrial control: theory and practice. Ann. Rev. Control **37**(1), 69–88 (2013)
14. Mahdjoub, M., Monticolo, D., Gomes, S., Sagot, J.C.: A collaborative design for usability approach supported by virtual reality and a multi-agent system embedded in a PLM environment. Comput. Aided Des. **42**(5), 402–413 (2010)
15. Wu, D., Rosen, D.W., Wang, L., Schaefer, D.: Cloud-based design and manufacturing: a new paradigm in digital manufacturing and design innovation. Comput. Aided Des. **59**, 1–14 (2014)
16. Wang, L.: Machine availability monitoring and machining process planning towards cloud manufacturing. CIRP J. Manufact. Sci. Technol. **6**(4), 263–273 (2013)
17. Morariu, O., Morariu, C., Borangiu, T., Raileanu, S.: Smart resource allocations for highly adaptive private cloud systems. J. Control Eng. Appl. Inform. **16**(3), 23–34 (2014)
18. Kärkkäinen, M., Ala-Risku, T., Främling, K.: The product centric approach: a solution to supply network information management problems? Comput. Ind. **52**(2), 147–159 (2003)

19. Vermesan, O., Friess, P.: Internet of Things—From Research and Innovation to Market Deployment. River Publishers (2014)
20. Dini, P., Lombardo, G., Mansell, R., Razavi, A., Moschoyiannis, S., Krause, P., Nicolai, A.L. R.: Beyond interoperability to digital ecosystems: regional innovation and socio-economic development led by SMEs. International Journal of Technological. Learning 1, 410–426 (2008)
21. Wu, D., Thames, J.L., Rosen, D.W., Schaefer, D.: Enhancing the product realization process with cloud-based design and manufacturing systems. J. Comput. Inf. Sci. Engi. 13(4) (2013)
22. Mezgár, I., Rauschecker, U.: The challenge of networked enterprises for cloud computing interoperability. Comput. Ind. 65(4), 657–674 (2014)
23. Främling, K., Holmström, J., Loukkola, J., Nyman, J., Kaustell, A.: Sustainable PLM through intelligent products. Eng. Appl. Artif. Intell. 26(2), 789–799 (2013)
24. Ashton, K.: Internet things—MIT, embedded technology and the next internet revolution. In: Baltic Conventions, The Commonwealth Conference and Events Centre, London (2000)
25. Kärkkäinen, M., Holmström, J., Främling, K., Artto, K.: Intelligent products–a step towards a more effective project delivery chain. Comput. Ind. 50(2), 141–151 (2003)
26. Kubler, S., Främling, K., Derigent, W.: P2P data synchronization for product lifecycle management. Comput. Ind. 66, 82–98 (2015)
27. Czajkiewicz, Z.: Direct digital manufacturing—new product development and production technology. Econ. Organ. Enterp. 2(2), 29–37 (2008)

Integrated Scheduling for Make-to-Order Multi-factory Manufacturing: An Agent-Based Cloud-Assisted Approach

Iman Badr

Abstract The fourth revolution currently envisioned for manufacturing is characterized by the interconnection of distributed manufacturing facilities and the provision of their services through cloud computing. Customers will be allowed to customize products in a make-to-order strategy and select among the available facilities and services. This paper addresses the integrated scheduling of customer orders for a multi-factory, make-to-order manufacturing environment. Distributed facilities are represented as autonomous agents that generate the schedule through goal-oriented negotiations. Scheduling agents are abstracted from facilities-related data, which are made available along with auxiliary scheduling tools in the cloud. In this way, the proposed scheduling provides a generic solution that generates efficient schedules flexibly.

Keywords Integrated scheduling · Agent-based scheduling · Cloud computing · Industry 4.0

1 Introduction

Presently, manufacturing is undergoing a fourth revolution, the so called industry 4.0, characterized by higher integration of resources, enterprises and customers. This revolution is expected to achieve higher customer satisfaction at lower prices through the make-to-order or mass customization strategy. Every customer will be able to configure or even design his own product and have it delivered in a pre-specified time. The production flow of every order, starting from the collection of material through the processing of the parts to be assembled and up to shipping and delivery, may span geographically scattered facilities. The integration of these distributed facilities and resources is enabled by new communication technologies that bring about new models of ubiquitous computing such as cloud computing [1].

I. Badr (✉)
Science Faculty, Helwan University, Cairo, Egypt
e-mail: imanb@aucegypt.edu

© Springer International Publishing Switzerland 2016
T. Borangiu et al. (eds.), *Service Orientation in Holonic and Multi-Agent Manufacturing*, Studies in Computational Intelligence 640,
DOI 10.1007/978-3-319-30337-6_25

277

Undoubtedly, this revolutionary change in manufacturing is faced with the challenge of managing the distributed facilities and providing seamless integration of the manufacturing tasks. Manufacturing scheduling represents one of the inherently challenging tasks, which has been receiving much interest in academia and industry due to its complexity as an optimization problem coupled with its impact in gaining the lead in a highly competitive market. Scheduling is concerned with optimizing the timely allocation of resources to competing jobs, according to a specified set of criteria such as maximizing the throughput. Despite the richness of the scheduling literature throughout the decades, its impact in industry has been minimal mainly due to the confined view of scheduling as only concerned with machines and isolated from the dynamic changes of the environment [2]. Accordingly, scheduling remains a major obstacle in adopting the envisioned manufacturing model.

In this paper, an integrated view of distributed scheduling is adopted and a multi-agent cloud-assisted architecture for ubiquitous manufacturing environments is presented. The rest of the paper is organized as follows. Section 2 presents a brief review of literature. Section 3 analyses the problem and deduces the influencing factors. Section 4 overviews the proposed architecture. Section 5 describes the scheduling method. Section 6 concludes the paper with a summary and future work.

2 Literature Review

Recently, some research work has been proposed to tackle the problem of distributed product management in general and scheduling in particular. In [3, 4], cloud-assisted platforms for managing distributed manufacturing and employing a service-oriented paradigm are presented. In [5, 6], RFID technology is employed to deal with the dynamic production scheduling problem by capturing and analysing real-time data. The authors in [5] apply a Monte Carlo simulation to generate a production schedule based on the captured data. In [6], the main scheduling decisions are taken by the human managers at the different production stages.

In [7], the scheduling of production logistics is focused on and a system based on Internet of Things (IoT) and cloud computing for solving this problem is presented. On the other hand, Zhang et al. [8] focus on the production scheduling problem and propose a multi-agent based architecture for a ubiquitous manufacturing environment. A method based on Genetic Algorithms is employed to solve the machine scheduling problem. Sun et al. [9] study the integrated production scheduling and distribution problem for multi-factory with both inland transportation and overseas shipment. The proposed solution is based on a two-level fuzzy guided genetic algorithm.

3   Problem Analysis

The scheduling problem studied in this research is concerned with a make-to order manufacturing environment encompassing distributed factories and multiple warehouses, material suppliers and transportation companies. The integration of the distributed facilities takes place via a cloud that provides a ubiquitous access to customers and manufacturing stakeholders. As depicted in Fig. 1, customers are involved as active stakeholders in the manufacturing process. They are allowed to place orders for their customized products, which may be processed at scattered locations and finally assembled and delivered to their customers in a specified deadline.

Scheduling customer requests in such a distributed, ubiquitous and dynamic environment have to be made under consideration of the current status of the entire set of the involved entities. The production flow of distributed manufacturing may be summarized in the steps captured in Fig. 2. First, raw material is collected either from an internal or external warehouse. While in the former case no transportation is required, the collected material has to be transported to the shop floor of the factory in the latter case. This justifies the existence of an arrow by passing the

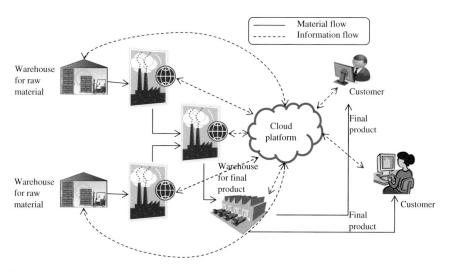

Fig. 1 A schematic illustration of distributed cloud-assisted manufacturing

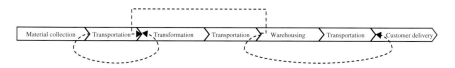

Fig. 2 Steps of the production flow of distributed manufacturing

transportation step to indicate that this step is optional. Followed by the collection and possible transportation of material, a transformation step takes place. This step involves a set of processing and material handling steps inside a factory.

The product that undergoes a transformation step may require further transformations at other factories or may correspond to the final product ordered by the customer. This is designated by the feedback loop depicted in Fig. 2, after the transportation step succeeding the transformation step. Taking the path, denoted by the feedback arrow, indicates the need to apply further transformations in other factories. Once a final product is produced, two possibilities exist, either directly transporting the product to deliver it to the customer or transporting it to be stored for some time before being transported again for customer delivery. While the former possibility corresponds to the path from the transportation step through the dashed arrow to the customer delivery step, the latter possibility is denoted by taking the rest of the steps (i.e. warehousing, transportation and customer delivery).

Each of these steps involves a set of resources, as captured by Table 1. To reduce the complexity associated with tracking this overwhelming set of resources, the influencing location or entity is identified for every step to be modelled later as an agent undertaking the allocation decision for the corresponding step. For example, the availability of raw material affects the material collection step and real-time tracking of material may be performed by attaching RFID to the material units. However, the allocation decision of material should be delegated to the material supplier rather than to the material itself. The transformation of material or work pieces inside a factory corresponds to a set of processing and material handling steps that are analysed and modelled as agents in [10, 11]. In multi-factory environment, every factory is responsible for its internal schedule and is conceived as the influencing entity in this case (see Table 1).

Integrated scheduling is affected by two facilities-related factors:

- *Static factors* related to the inherent specifications of facilities such as the maximum speed of a truck, the services supported by a factory and the capacity of a warehouse.
- *Dynamic factors* corresponding to the current status of facilities such as the current free space in a warehouse, the currently existing material at a certain store and the shipping capacity of a certain shipping company for a given date.

Table 1 A derivation of the influencing resources and entities of a typical material flow

Material flow step	Influencing resource	Influencing location/entity
Material collection	Material	Warehouse/material supplier company
Transportation	Trucks, forklifts, etc.	Transportation company
Transformation	CNCs, robots, AGVs, etc.	Factory
Warehousing	Location and internal facilities	Warehousing company

4   Agent-Based Scheduling Architecture

This work builds on a previously proposed agent-based architecture for a single factory with flexible resources [11]. By definition, scheduling is an intermediary task catering for the allocation of existing resources to the planned jobs. This is reflected in the automation pyramid which captures scheduling as an intermediary layer between planning and shop floor control. The agent-based architecture, which is extended in this work, decomposes scheduling autonomous agents at four layers representing a mapping of planning and shop floor control. Agents at the upper most layers represent customers and jobs and aim at optimizing the allocation of the set of jobs ordered by the corresponding. The two other layers model shop floor control and contain agents representing resources and shared services, at the lower most layer and the one above it respectively. Schedule generation and update result from the goal-oriented negotiations among the concerned agents, which drastically reduces the complexity compared to the conventional centralized approach. A near optimal solution with good reactivity is guaranteed through the heterarchical optimization in a way that capitalizes on the redundant capabilities of the flexible resources at the shop floor.

These agents are complemented with a centralized model that includes relevant details from the scheduling environment, namely planning and control. These details include raw material required for every part type, the technological order of operations required by every planned product and the operations supported by every machine. This rationale behind the environmental model is to provide a generic support for scheduling which greatly facilitates the adoption and extensibility. For an elaborate discussion of the architecture, the reader may refer to [11].

As illustrated in Fig. 3, the discussed architecture is extended to the multi-factory make-to-order manufacturing by first adding a "facilities" layer in between the previously explained four layers. This additional facilities layer accounts for the distributed nature of the problem and integrates all entities involved in the entire production flow. This integration becomes possible by defining agents representing the influencing entities, derived in the previous section. Four agent types represent

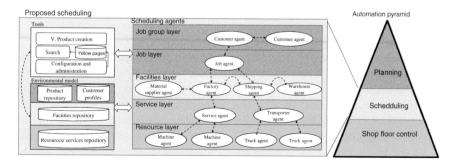

Fig. 3 The proposed agent-based cloud-assisted architecture

the four entities, listed in Table 1, namely a *supplier* agent, a *shipping* agent, a *factory* agent and a *warehouse* agent.

Similarly, the environmental model is extended by incorporating the static factors defined in the previous section in a facilities repository. Furthermore, the static factors derived from the resources and services of all the facilities have to be added to the existing resources and services repositories, respectively. To account for the customer-oriented production and make-to-order strategy, customer profiles should be added as well. Each profile captures personal information, including address or delivery location, preferences and a history of designed, configured and ordered products.

Auxiliary tools are required to populate the environmental model with facilities, products, services, etc. This is made possible through configuration and administration tools that provide enabling the registration and deletion of entities within the environmental model. Search tools are also provided to allow agents to find each other dynamically based on the provided service. To enable customers to make their own products, a tool for the design and creation of virtual products has to be provided as well.

5   Scheduling Method

The schedule generation takes place dynamically through the goal-oriented negotiations among the concerned agents. The proposed scheduling method may be summarized in the following steps.

- Initializations
  Before placing an order, the customer navigates through a product repository and either selects a predefined product, configures an existing one or designs a new product. A customer order corresponds to a job for manufacturing a specific product in a given quantity, at a delivery time or deadline and possibly a maximum price. The customer agent reacts to the placement of an order by instantiating a job agent. The job agent retrieves the technological order of the product in concern, i.e. the production services required to manufacture the product along with their sequence. The factory agents supporting these services are retrieved and contacted.
- Collecting production scheduling proposals
  To generate a production schedule, the factory agents supporting the required services are contacted by the job agent. Every factory agent prepares and sends its bid to the job agent. A bid includes the earliest start time, the latest end time, and an estimated price. In case raw material or parts are required from an external warehouse, the corresponding supplier agent is retrieved from the yellow pages through the search tool. The cost and time of the supply is considered by the factory agent when generating its bid. In generating a bid, factory agents decompose the required service into internal services supported by the

internal workstations. The optimization of bids at each factory is performed at the service layer which represents the shared services of the internal workstations. Incorporating search heuristics such as Genetic Algorithms in this two-layer optimization provides a good combination between efficiency and flexibility, as proved in a previous study [12].

- Collecting material handling proposals
  The job agent attempts in this step to find material handling proposals that provide the least cost and meet the required delivery time. The shipping of unfinished/finished parts from a source factory to a destination factory or to the customer delivery point is sought by contacting concerned shipping agents. A shipping agent is retrieved when its covered map spans the locations of source and destination. It replies to job agents' requests by a proposal consisting of the date and time of both collection and delivery from source to destination. In arranging for shipping from source to destination, planned production schedule at the concerned factories is compared to the proposed shipping times to decide whether temporary storage is needed. If temporary storage is needed, shipping agents contact warehouse agents corresponding to warehouses that are located nearby to reserve storage space. The cost and time of both the temporary storage and the shipping are incorporated in one proposal by the shipping agent and sent back to the job agent.
- Evaluating proposals and committing schedule
  All possible combinations of collected proposals both from the factories agents as well as the shipping agents are evaluated in terms of cost and time. The best proposal is then selected and a confirmation is sent to the concerned agents to commit the integrated schedule.

6   Conclusion

The present work deals with the integrated scheduling problem for a distributed multi-factory, make-to-order manufacturing. The problem is analysed to identify the involved entities along with static and dynamic characteristics of these entities influencing the flow of production. The influencing entities are modelled as autonomous agents and added to an agent-based, single-factory architecture proposed in a previous work. The extended architecture accommodates for the new agents at an additional layer of abstraction representing the distributed manufacturing facilities that spans the entire flow of production starting from material collection up to customer delivery. The architecture encompasses a total of five layers of abstraction that contain agents which encapsulate and capture the dynamic influencing factors of the corresponding entities. The agents are complemented with an environmental model that keeps track of the static factors, influencing the flow of production and a set of auxiliary tools that are used for configuring and customizing products, registering and deleting services and agent, searching for services and agents, etc.

The generation of an integrated schedule takes place through the dynamic goal-oriented negotiations among the concerned agents. By basing schedule decisions on the factors influencing the flow of production and captured dynamically, agents can better react to unforeseen events. The decomposition of the scheduling problem into agents and limiting the negotiations among only the concerned agents greatly reduce the time complexity and thus enhances the responsiveness. Furthermore, the incorporation of static influencing factors exogenous to scheduling agents facilitates the adoption of the proposed scheduling regardless of the specifications of the involved factories.

Work is ongoing on employing different optimization algorithms that can be incorporated in the proposed agents. The different algorithms are to be evaluated on different case studies.

References

1. Zhang, Q., Cheng, L., Boutaba, R.: Cloud computing: state-of-the-art and research challenges. J Internet Serv. Apdpl. **1**, 7–18 (2010)
2. Ouelhadj, D., Petrovic, S.: Survey of dynamic scheduling in manufacturing systems. J. Sched. **12**(4), 417–431 (2009)
3. Huang, B., Li, C., Yin, C., Zhao, X.: Cloud manufacturing service platform for small- and medium-sized enterprises. Int. J. Adv. Manuf. Technol. **65**, 1261–1272 (2013)
4. Valilai, O., Houshmand, M.: A collaborative and integrated platform to support distributed manufacturing system using a service-oriented approach based on cloud computing paradigm. Robot. Comput. Integr. Manuf. **29**, 110–127 (2013)
5. Guo, Z., Yang, C.: Development of production tracking and scheduling system: a cloud based architecture. In: International Conference on Cloud Computing and Big Data (2013)
6. Luo, H., Fang, J., Huang, G.: Real-time scheduling for hybrid flow shop in ubiquitous manufacturing environment. Comput. Ind. Eng. **84**, 12–23 (2015)
7. Qu, T., Lei, S., Wang, Z., Nie, D., Chen, X., Huang, G.: IoT-based real-time production logistics synchronization system under smart cloud manufacturing. Int. J. Adv. Manuf. Technol. 1–18 (2015)
8. Zhang, Y., Huang, G., Sun, S., Yang, T.: Multi-agent based real-time production scheduling method for radio frequency identification enabled ubiquitous shop floor environment. Int. J. Comput. Ind. Eng. **76**, 89–97 (2014)
9. Sun, X., Chung, S., Chan, F.: Integrated scheduling of a multi-product multi-factory manufacturing system with maritime transport limits. Transp. Res. Part E Logistics Transp. Rev. **79**, 110–127 (2015)
10. Badr, I., Göhner, P.: An agent-based approach for scheduling under consideration of the influencing factors in FMS. In: The 35th Annual Conference of the IEEE Industrial Electronics Society (IECON-09). Porto, Portugal (2009)
11. Badr, I.: Agent-based dynamic scheduling for flexible manufacturing systems. Dissertation Thesis (2010)
12. Badr, I., Göhner, P.: Incorporating GA-based optimization into a multi-agent architecture for FMS scheduling. In: the 10th IFAC Workshop on Intelligent Manufacturing Systems. Lisbon, Portugal (2010)

Secure and Resilient Manufacturing Operations Inspired by Software-Defined Networking

Radu F. Babiceanu and Remzi Seker

Abstract Software-Defined Networking (SDN) is a relatively new concept in the cloud and computer networks domain proposed as a solution to depart from the current limitations of traditional IP networks which are complex and, many times, difficult to manage. Manufacturing operations, being them originated on a single shop-floor or distributed across many organizations, have long now been subject to limitations in performance due to the manufacturing control software. This paper investigates the SDN concept adoption for the manufacturing product design and operational flow, by promoting the logical-only centralization of the shop-floor operations control within the manufacturing shared-cloud for clusters of manufacturing networks. First, the paper proposes the adoption of SDN concept to distributed manufacturing networks, with the goal to improve the performance of manufacturing data network metrics. Then, the paper proposes the design of an SDN-inspired mechanism for manufacturing control, with the goal to optimize the performance of specific manufacturing operations metrics such as total completion time, maximum lateness, and others. Both solutions are expected to bring manufacturing operations similar benefits that SDN is reported to generate to IP-based networks.

Keywords Cloud manufacturing · Software-defined networking · Manufacturing logical control

R.F. Babiceanu (✉) · R. Seker
Department of Electrical, Computer, Software, and Systems Engineering, Embry-Riddle Aeronautical University, Daytona Beach, FL 32114, USA
e-mail: babicear@erau.edu

R. Seker
e-mail: sekerr@erau.edu

© Springer International Publishing Switzerland 2016
T. Borangiu et al. (eds.), *Service Orientation in Holonic and Multi-Agent Manufacturing*, Studies in Computational Intelligence 640,
DOI 10.1007/978-3-319-30337-6_26

285

1 Introduction

Software-Defined Networking (SDN) is a relatively new concept in the computer networks domain that was proposed as a solution to depart from the current limitations of traditional IP networks which are complex and, many times, difficult to manage [1]. Internet IP network administrators have to work with predefined policies for network management, which makes it hard to reconfigure the network to respond to faults, heavy network loads, or unexpected changes. Moreover, current IP networks are also vertically integrated, with the control plane, responsible for network traffic, and the data plane, responsible for forwarding the traffic based on the decisions made at the control plane level, residing on the actual network devices, such as routers and switches, which makes the network management even harder, at times [2]. SDN concept separates the vertical integration of the control and data planes and allows the control software to be executed in the cloud or designated IT servers, thus, removing the control overhead from the network devices, which in this new framework will only be responsible for packet forwarding [3]. The SDN control function is implemented by a logically centralized controller, which benefits from the global view of the network and makes informed decisions in relation to the network load and traffic.

Manufacturing operations, being them originated on a single shop-floor or distributed across many organizations, have long now been subject to limitations in performance due to the manufacturing control software. Centralized manufacturing control is not flexible and optimal solutions for routing and scheduling manufacturing jobs can only be derived if the optimization problem is small [4]. Since manufacturing sequencing and scheduling is well-known to be a NP-hard problem, practitioners relied for decades on heuristics. On the other side of the spectrum, decentralized manufacturing scheduling offers flexibility and is reactive to order changes or resource failures, but given its intrinsic decentralized nature, it cannot guarantee an optimal solution [5]. In practice, decentralized routing and scheduling solutions are fast and perform well, unless the algorithm gets stuck in a local optimum, in which case the solution offered is fouled by the lack of global view of the system. Moreover, the advances in sensor and communication technologies, and the prevalence of cyber-physical devices forming the Internet-of-Things (IoT), can provide the foundations for linking the physical manufacturing facilities world to the cyber world of software and Internet applications [6].

This paper investigates the adoption of the SDN concept into the manufacturing product design and operational flow, by promoting the logical centralization of the shop-floor operations control within the manufacturing shared-cloud for clusters of manufacturing networks. First, the paper proposes the adoption of SDN concept to distributed manufacturing operations, with the goal to improve the performance of the manufacturing data network metrics. This solution not tested previously in manufacturing environments is expected to bring manufacturing operations similar benefits that SDN is reported to bring to IP networks, such as reduced data packet delays, corruption, and loss, and increased security and resilience of data packet transmission process. Also, this work may drive the manufacturing research and practitioners communities to speed-up the adoption of SDN to distributed manufacturing operations. Secondly, the paper proposes the design of a novel SDN-inspired control mechanism for manufacturing sequencing and scheduling, with the goal to optimize the performance of specific manufacturing operations metrics such as total completion time, maximum lateness, and others. This novel solution is expected to bring to manufacturing control similar benefits that SDN is reported to bring to IP networks, such as optimized manufacturing resource routing solutions, better resource load balancing, and improved monitoring (fault-detection) of manufacturing resources.

This work may also provide the cloud manufacturing research community with the foundations to tackle complex optimization problems, which many times resort on heuristics for acceptable solutions. Moreover, cloud control of manufacturing decisions will come with the added benefit of the cyber security solutions that cloud platforms offer. From this point forward the paper is structured as follows: Sect. 2 provides a review of the most important aspects of the SDN and manufacturing control, and after that, Sect. 3 presents the proposed Manufacturing-SDN System model, detailing certain critical modelling aspects and instantiates the manufacturing-SDN systems paradigm. Finally, the future research concerning the proposed system framework is outlined in the conclusions section.

2 Literature Review

2.1 Brief Review on Software-Defined Networking

The Open Networking Foundation (ONF), a non-profit consortium dedicated to software-defined networking development, standardization, and commercialization, defines SDN as "an emerging network architecture where network control is decoupled from forwarding and is directly programmable" [7]. Through this decoupling of the control logic from the network hardware (routers and switches), which become in this case simple forwarding devices, advantages such as consolidation of middle boxes, simpler policy management, and new functionalities are obtained [8]. The main advantage of employing different data and control planes is that the forwarding decisions are changed from being destination-based solutions resulted from local system load and traffic, to logically centralized flow-based decisions, where a flow is a sequence of packets between a source and a destination.

The separation between the data plan and the control plan is made by an application programming interface (API), the most used in current SDN implementations being OpenFlow [7]. The OpenFlow API links the routers and switches of the data plane with the SDN controller residing on the control plane [3]. These characteristics of the SDN networks are shown in Fig. 1. The SDN controllers are also depicted.

Fig. 1 SDN separation of
data and control planes

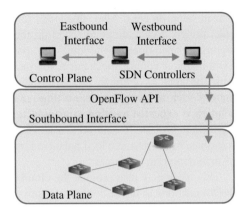

While solutions that include a single controller may be sufficient to manage small networks, they are not scalable for large networks and also may represent single point of failures. From this reason, cluster of controllers are used in order to achieve a higher degree of resiliency and for supporting a larger number of network devices. Solutions that include entire Network Operating Systems (NOS), which facilitate the programmability of the network, are considered, as well. SDN controllers or NOS perform forwarding table changes by adding, removing and updating forwarding tables on the network devices [9].

2.2 Brief Review on Cloud Manufacturing Control

Traditional solutions for manufacturing control range from total centralized to total decentralized control schemes. Manufacturing control literature presented over the years the advantages and disadvantages of both and any other solutions in between [4]. More recently, the revolutionary adoption of virtualization and cloud computation at different level of service (IaaS, PaaS, HaaS, SaaS) in many areas was embraced by some well-established manufacturing organizations, as well. Other manufacturing organizations acting as service broker were established and operate in the overall virtualized and cloud manufacturing environment [3].

Previous authors' work [10] proposes a Manufacturing Cyber-Physical System (M-CPS) model, which includes both the physical world, where the traditional manufacturing system is located, and the cyber world, where the Internet connectivity and computing in the cloud is performed. Figure 2 presents the proposed M-CPS system.

In between the two worlds, there is a layer of cyber-physical devices, such as sensors and actuators, local area networks, and also application and cyber security software, which completes the cyber-physical system model. The layer of

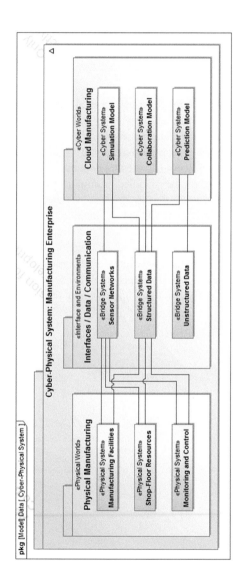

Fig. 2 The manufacturing cyber-physical system [10]

cyber-physical devices, when properly deployed with the needed redundancy, is able to provide status control through the sensors and provide adjustments to any stages of the manufacturing operations through the actuators.

3 The Proposed Manufacturing-SDN Model

3.1 Proposed Manufacturing-SDN System Framework

The proposed Manufacturing-SDN (M-SDN) system model is depicted in Fig. 3 below, together with the correspondent pure SDN and testbed simulation modules.

In traditional IP networks the data and control plan are tightly integrated and located in the network devices, such as routers and switches scattered across the network. The entire routing process is decentralized, and the forwarding decisions are made by the routers based on the so-called routing algorithms. Similar decisions are made in decentralized manufacturing systems, where lower level controllers distribute among themselves the workload of incoming orders, based on their capabilities and existing and scheduled workload, without any overall global view. While this type of order routing and scheduling decisions provides on-time good solutions, there still exists the risk that the individual algorithms run at the resource controller lower level result in less than acceptable performance. Getting stuck in local optimum solutions, when running optimization algorithms, could be one

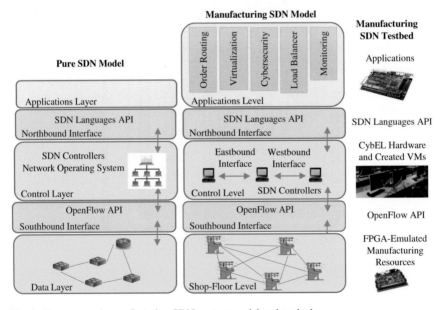

Fig. 3 The proposed manufacturing-SDN system model and testbed

potential problem, with resulting undesired consequences from the delivery time
and cost points of view. Relying on solutions provided by dispatching rules is
another problematic approach, as none of those simple heuristic rules will provide
acceptable solutions every time they are called.

As mentioned above, SDN separates the data and control plans in IP networks
with the benefits of better allocation of load in the network and control over the
route to be followed by the IP data packets from source to destination, among
others. Similar benefits can be obtained in the case of adopting SDN concept to
manufacturing operations. The adoption approach essentially needs to start with a
distributed manufacturing architecture as centralized architectures do not have the
flexibility in communication needed at lower levels of the architecture for forwarding manufacturing packets across the shop-floor network of machines.

3.2 Adoption of SDN Concept to Distributed Manufacturing Networks

The first part of this work is the adoption of the SDN concept to distributed
manufacturing operations, with the goal to improve the performance of the manufacturing data network metrics. As depicted in Fig. 3, the M-SDN testbed uses
FPGA boards to emulate the manufacturing resources data on the shop-floor level.
The proposed testbed includes several FPGA Development Boards, such as the
powerful Xilinx Zynq-7000 Series featuring the Zynq All Programmable
System-on-Chip, which can be used to implement OpenFlow devices yielding
88Gbps throughput for 1 K flow supporting dynamic updates [11]. Current available OpenFlow devices are able to provide up to 1000 K forwarding table entries
for the Southbound Interface model. However, the testbed will use Open vSwitch, a
virtual OpenFlow device software that provides a number of virtual data points
larger than the actual hardware ones. Open vSwitch is available from their portal
[12]. The testbed will use two solutions for the Control Level modelling purpose. In
a first step, Mininet, a platform available from Mininet portal [13], will be considered. Mininet creates a realistic virtual network, running real kernel, switch and
application code on a single machine which could be VM, cloud or native. One of
the key properties of Mininet is its use of software-based OpenFlow switches in
virtualized containers, providing the exact same semantics of hardware-based
OpenFlow switches [3]. Once testing is successful, the testbed will employ the use
of multiple VM created in a computer network environment.

The Northbound Interface will be emulated by employing SDN programming
languages such as Procera, NetCore, Pyretic, etc. The Application Level will
include at a minimum network applications for load balancing, virtualization, cyber
security, process monitoring, and manufacturing control routing (manufacturing
order sequencing and scheduling). The proposed implementation, depicted in
Fig. 4, will be loaded with manufacturing orders to be performed in different

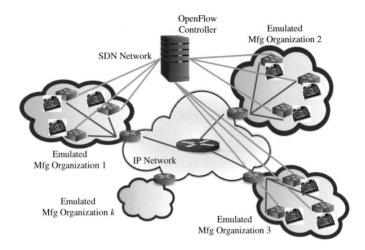

Fig. 4 The proposed implementation of manufacturing-SDN (Adapted from [2])

sections of the overall network, mimicking geographically distributed operations with the imposed need that the data files travel across the virtual networks, where data packets can be delayed, lost, or corrupted [2]. The same scenarios will be loaded to study the differences in the performance and quality factors when employing SDN.

3.3 Design of an SDN-Inspired Mechanism for Manufacturing Control

The second part of this work is the design of a novel SDN-inspired control mechanism for manufacturing sequencing and scheduling, with the goal to optimize the performance of specific manufacturing operations metrics such as total completion time, maximum lateness, etc. While the actual manufacturing orders are still residing in the data plan, the decision on where they are moved next is not made at the resource level as in the decentralized manufacturing control, but it will be done at the control plane, where the global view of the network obtained through the logically centralized control will help in providing solutions that avoid the limitations of the decentralized control. The first scenarios tested will include previously solved agent-based and holonic approaches solutions reported in the area of decentralized manufacturing control [4, 5, 14]. Scenarios will increase in complexity to the level of what is expected to be realistic for today's cyber-physical systems that include large numbers of systems and components that need to be manufactured and assembled together in different locations geographically distributed before a final assembly operation is performed. Requirements analysis, engineering design and analysis, manufacturing operations, quality assurance, cyber security data, imaging, application, and executable files will be needed at minimum during such an engineering effort.

3.4 Security and Resilience Analysis for the Manufacturing-SDN Model

SDN security and resilience are identified among the SDN research areas currently pursued both in the academia and industry. Clustered control architectures such as SDN that provide distributed functionality are under constant scrutiny for their availability and scalability and thus their resilience characteristics. Also, the SDN decoupling between the data and control layers can result in delays in reporting of faulty data links due to communication overload. Therefore, the resilience of SDN OpenFlow networks depends on both the fault-tolerance in the data layer, as well as the performance of the logically (only) centralized control layer. An SDN-inspired solution offers manufacturing the virtualization and cloud capabilities to address changes and respond to resource failures in practically real-time. Also, VM-enabled control of manufacturing decisions comes with the added benefit of cyber security solutions that cloud platforms offer.

4   Conclusions and Future Work Directions

This work presents a framework for the development of a secure and resilient Manufacturing-SDN system, where the cloud manufacturing operations are expected to embrace the benefits of deployed SDN in traditional IP networks. The work proposes an actual Manufacturing-SDN testbed and outlines the components and test scenarios for the proposed system. Future work will address the testbed implementation and report on the results of the SDN distributed cloud manufacturing operations.

References

1. Xia, W., Wen, Y., Xie, H., Foh, C.H., Niyato, D., Xie, H.: A Survey on Software-defined networking. IEEE Commun. Survey Tutorials **17**(1), 27–51 (2015)
2. Hakiri, A., Gokhale, A., Berthou, P., Schimdt, D.C., Gayraud, T.: Software-defined networking: challenges and research opportunities for future internet. Comput. Netw. **75**, 453–471 (2014)
3. Kreutz, D., Ramos, F.M.V., Verissiomo, P.E., Rothenberg, C.E., Azodolmolky, S., Uhlig, S.: Software-Defined networking: a comprehensive survey. Proc. IEEE **103**(1), 14–76 (2015)

4. Babiceanu, R.F., Chen, F.F.: Development and applications of holonic manufacturing systems: a survey. J. Intell. Manuf. **17**(1), 111–131 (2006)
5. Babiceanu, R.F., Chen, F.F.: Distributed and centralized material handling scheduling: comparison and results of a simulation study. Rob. Comput. Integrated Manuf. **25**(2), 441–448 (2009)
6. Babiceanu, R.F., Seker, R.: manufacturing operations, internet of things, and big data: towards predictive manufacturing systems. In: Borangiu, T., Thomas, A., Trentesaux, D. (eds.) Service Orientation in Holonic and Multi-Agent Manufacturing, SCI, vol. 594, pp. 157–164. Springer, Heidelberg (2015)
7. Open networking foundation. https://www.opennetworking.org
8. Nunes, B.A., Mendonca, M., Hguyen, X.-N., Obraczka, K., Turletti, T.: A survey of software-defined networking: past, present, and future of programmable networks. IEEE Commun. Survey Tutorials **16**(3), 1617–1634 (2014)
9. Jarraya, Y., Madi, T., Debbabi, M.: A survey and a layered taxonomy of software-defined networking. IEEE Commun. Survey Tutorials **16**(4), 1955–1982 (2014)
10. Babiceanu, R.F., Seker, R.: Manufacturing cyber-physical systems enabled by complex event processing and big data environments: a framework for development. In: Borangiu, T., Thomas, A., Trentesaux, D. (eds.) Service Orientation in Holonic and Multi-Agent Manufacturing, SCI, vol. 594, pp. 165–173. Springer, Heidelberg (2015)
11. Kobayashi, M., Seetharaman, S., Paruklar, G., Appenzellar, G., Little, J., van Reijendam, J., Weismann, P., McKeown, N.: Maturing of OpenFlow and software-defined networking through deployments. Comput. Netw. **61**, 151–175 (2014)
12. Open vSwitch. http://vswitch.org
13. Mininet. http://mininet.org
14. Mejjaouli, S., Babiceanu, R.F.: Holonic condition monitoring and fault-recovery system for sustainable manufacturing enterprises. In: Borangiu, T., Thomas, A., Trentesaux, D. (eds.) Service Orientation in Holonic and Multi-Agent Manufacturing, SCI, vol. 544, pp. 31–46. Springer, Heidelberg (2014)

Building a Robotic Cyber-Physical Production Component

Paulo Leitão and José Barbosa

Abstract Cyber-physical systems are a network of integrated computational decisional components and physical elements. The integration of computational decisional components with the heterogeneous physical automation systems and devices is not transparent and constitutes a critical challenge for the success of this approach. The objective of the paper is to describe an approach to establish standard interfaces based on the use of the ISO 9506 Manufacturing Message Specification international standard. The proposed approach is exemplified by the construction of a robotic cyber-physical production component that is plug-in in a cyber-physical system for a small-scale production system based on Fischertechnik systems.

Keywords Cyber-physical systems · Industrial agents Integration

1 Introduction

The manufacturing world is being subject to a paradigm shift, both at the organizational and control levels, facing the current demands for more robust, flexible, modular, adaptive and responsive systems. Several opportunities arise for the introduction of new and innovative approaches, such as the Cyber-Physical System (CPS) approach. This CPS approach is being supported by strong financing

P. Leitão · J. Barbosa (✉)
Polytechnic Institute of Bragança, Campus Sta Apolónia,
5301-857 Bragança, Portugal
e-mail: jbarbosa@ipb.pt

P. Leitão
e-mail: pleitao@ipb.pt

P. Leitão
LIACC—Artificial Intelligence and Computer Science Laboratory,
Rua Dr. Roberto Frias, 4200-465 Porto, Portugal

© Springer International Publishing Switzerland 2016
T. Borangiu et al. (eds.), *Service Orientation in Holonic and Multi-Agent Manufacturing*, Studies in Computational Intelligence 640,
DOI 10.1007/978-3-319-30337-6_27

295

measures, such as the European Horizon 2020 framework or the German Industry 4.0 initiative, leveraging a new industrial revolution and capturing the attention of academia or industry.

CPS constitutes a network of interacting cyber and physical elements aiming a common goal [1]. A major challenge is to integrate the computational decisional components (i.e. cyber part) with the physical automation systems and devices (i.e. physical part) to create such network of smart cyber-physical components. However, this integration is not transparent and constitutes a critical challenge for the success of this approach. In fact, it is not easy and transparent to integrate heterogeneous automation devices, such as sensors, robots, numerical control machines or automation solutions based on Programmable logic Controllers (PLCs), which usually requires a complex and time consuming activity. To face this problem, the challenge is to define standard industrial interfaces that allow a completely transparent development of the computational decisional components without knowing the particularities of the automation device; in such process, these interfaces may be developed by automation providers or system integrators and (re-)used by the system developers.

The objective of the paper is to describe an approach to establish such standard interfaces based on the use of the ISO 9506 Manufacturing Message Specification (MMS) international standard [2], initially introduced by ADACOR holonic control architecture [3]. This approach is exemplified by deploying a cyber-physical production component for an industrial manipulator robot, which is part of a small-scale production system based on Fischertechnik systems.

The rest of the paper is organized as follows: Sect. 2 overviews the concept of cyber-physical systems and identifies the integration of computational components with automation devices as a critical challenge for its industrial implementation. Section 3 presents an approach to engineer cyber-physical production components and Sect. 4 illustrates its applicability by developing a robotic cyber-physical production component. Finally, Sect. 5 rounds up the paper with the conclusions.

2 Overview of Cyber-Physical Systems

Embedded systems have been in use for many years. They can be characterized by the conjunction of computational, electrical and mechanical capabilities, being often executed in real-time and providing some sort of intelligence to the system. Embedded systems are present everywhere and in different sectors, such as civil infrastructure, aerospace, energy, healthcare, manufacturing, transportation. Examples are vendor machines, cars' Automatic Breaking System (ABS) or even elevators.

With the widely improvement and spread of communication technologies, namely wireless communication and optical fibre, used currently in internet

Fig. 1 Cyber-physical
system triad [4]

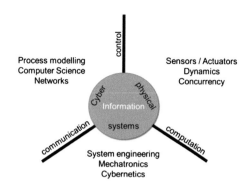

infrastructures, these embedded systems have gained a mean to share information, cooperate and collaborate with each other. This missing communication capability and the collaboration inability of embedded systems gave rise to CPS. In this way, CPS can be defined as a triad of computation capabilities with a control component, interconnected over a communication channel, as depicted in Fig. 1.

The cyber, or logic world, can be found at the upper level of this triad, being responsible for bringing the logic and intelligence features found in CPS. At the lower level resides the physical process, being often composed by the combination of electromechanical components. This triad is complete by the communication capability, tying every Cyber Physical Component (CPC) from which the CPS is built upon.

The application of CPS will impose tremendous changes at several levels, particularly in the way the systems are designed and how they interact. Referring to Fig. 1, at the cyber level, the challenges are related to process modelling, computer science and communication networks, while at the system level, new system engineering methodologies, developments in mechatronics and a new cybernetics discipline approach will be mandatory. At the physical level, the challenges are related to sensors and actuators, dynamics and concurrency. In manufacturing, CPS will imply the migration from the typical ISA-95 organizational structure, where all levels are vertically interconnected into a decentralization of these levels, meaning that the components/applications placed at different levels can access to data provided by others.

An important feature is the need to integrate the cyber part, i.e. computational decisional components, with the physical world, which are responsible to sense, process and act on the environment. This fundamental feature, particularly in manufacturing, implies several challenges where the definition of standardized interfaces assumes a crucial importance to handle the usually heterogeneous automation device presented at shop floor.

3 Engineering Cyber-Physical Components

The engineering of cyber-physical production components requires the integration of the computational and physical automation counterparts. The computational components may use the agent technology [5] to implement the intelligence and adaptation layer that will control the automation hardware (HW) device. Intelligent software agents developed in this context are known as industrial agents, which are faced with industrial requirements, namely HW integration, reliability, fault tolerance, scalability, industrial standards compliance, resilience, manageability, and maintainability [6]. Additionally, the integration of computational and physical automation counterparts recalls the *holon* concept, which is composed by an informational component (the agent) and the physical component (the HW device if exists) [7].

The integration of cyber and physical components can be performed in two different manners, namely embedding the agent within the physical control device or connecting the agent with the existing control device in a coupled manner [8, 9]. Independently of the use of these two approaches, it is necessary to create a standard approach that allows the transparent and independent development of the computational entities from the heterogeneity and particularities of the HW automation device and communication infra-structure, which can expose their functionalities in terms of services. This imposes a crucial challenge for the engineering of these components regarding the establishment of standard interfaces, focusing the semantics and the protocols, and industrial middleware.

Having this in mind, ADACOR holonic manufacturing control architecture [3] proposed an approach based on the use of standard interfaces using the service-oriented architectures (SoA) principles, where the physical automation resource is abstracted in form of standard services, as illustrated in Fig. 2. These services were defined based on the ISO 9506 Manufacturing Message Specification (MMS) standard [2], which defines the syntax and semantics for a set of clusters of services for automation domain, which are invoked by the agent (which plays the role of client) independently from the device particularities. MMS was designed as a control and monitoring specification for the OSI application layer, enabling the cooperation between applications and/or devices at the shop-floor. This specification is used in this approach to provide the necessary interface specification guidelines, namely allowing to define standard services that expose the functionalities of the HW automation device, namely in terms of variable handling, program handling and events [2].

These services are implemented, usually by system integrators in the server component according to the particularities of the device available at the shop-floor

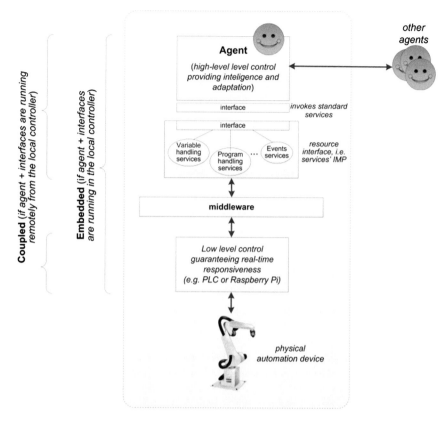

Fig. 2 Integration of agents with low level automation functions to form a cyber-physical production component

(from different types, e.g. robots and numerical control machines, and from different automation providers) and the communication infrastructure (e.g. serial communication, Modbus or OPC-UA). These services, after being developed, can be re-used and offered as drivers or wrappers, to be used in a pluggable and modular manner by other control applications for similar resources.

The transparent and standard invocation of these services by the computational entity requires the specification of the syntax of each service, i.e. the definition of input and output parameters. The services available in the *Program Invocation Service* package, which are invoked in a unique way by the client side, i.e. the agents, are:

```
public interface ProgramInvocationService {
    boolean CreateProgramInvocation(String program);
    boolean DeleteProgramInvocation(String program);
    ...
    boolean Start(String program);
    boolean Stop(String program);
    boolean Resume(String program);
    boolean Reset(String program);
    boolean Kill(String program);
    Attributes GetProgramInvocationAttributes(String program);
    boolean Select(String program);
    Attributes AlterProgramInvocationAttributes String program, At-
tributes att);
    ...
    boolean ReconfigureProgramInvocation(String program);
}
```

Note that, as an example, whenever an agent needs to start the execution of a robotic a program, it uses the service *Start (String program)*, where *program* represents the program to be executed.

At this point, agents invoke standardized services based on an abstraction layer, which enables the transparent design and development of agents. The challenge here focuses on the abstraction level that the standard interfaces impose to the generic development of agents. In fact, this abstraction culminates with the generalization of the parameters of the methods to be executed, where two instantiated agents of the same type have the same method invocation but in reality the physical access differs in each case. For instance, the parameter writing in a memory space differs from one agent to the other, accordingly with the HW to be accessed (e.g. the physical address where the temperature sensor is connected to the PLC).

As illustrated in Fig. 3, initially when launched, each agent parameterizes the generic parameters used in the MMS-based interface layer in a *xml* configuration file. This file contains pairs of *{tag, value}*, where *tag* represents the service parameter name used in the agent development, while *value* points to the

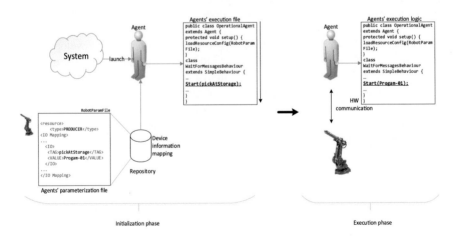

Fig. 3 Agent-HW interface: initialization and execution phases

physical/logical connection that the agent need to access. It is worthy to be noted also that *value* can represent more complex structures aside direct/simple type. On example of such pair is *{part_input, IRB1400.signaldi.DI10_1}*, where *part_input* represents the sensor that detects that a part is at the beginning of a conveyor, and *IRB1400.signaldi.DI10_1* is the physical label used in the OPC server. This is crucial to guarantee that two similar automation devices have the same processing logic from the agent's point of view but due to their HW differences they must be parameterized differently.

4   Deploying a Robotic Cyber-Physical Component

This section describes the application of the described engineering approach to deploy a robotic cyber-physical component to be used in an agent-based control system for a small scale production system.

4.1   Description of the Case Study

The robotic device is an IRB 1400 ABB robot that is part of a real small-scale production system, which also comprises two punching machines and two indexed lines supplied by Fischertechnik™, as illustrated in Fig. 4.

The punching and indexed machines are controlled by IEC 61131-3 programs running in a Modicon M340 PLC. Two different parts circulate in the system, each one having a particular process plan. The circulation of parts within the flexible production system is tracked by radio-frequency identification (RFID) readers. An industrial manipulator robot executes the transfer of the parts between the machines using proper RAPID programs and is accessible through the ABB S4 DDE Server (that can be accessed by OPC).

Fig. 4 Layout of the
small-scale production system

The idea in this work is to describe the way the cyber-physical production component for the industrial robot was engineered, and particularly how the software agent, which is providing intelligence and adaptation to the robot, was interfaced with the physical controller of the automation device.

4.2 Development of the Software Agent

An ADACOR-based system [3] was developed to control in a distributed manner this small-scale production system, using agent technology to implement the control logic, i.e. the cyber part. For this purpose, product, task, operational and supervisor holons were developed to represent the system components, each one contributing with their knowledge and skills to achieve the system's goals, by interacting through several co-operation patterns. Particularly, two product holons were created, one for each product type defined in the catalogue of available parts. Task holons are launched by the associated product holons according to the order demand. The ecosystem of heterogeneous resources has associated an operational holon (OH) for each one, i.e., for the two punching machines, the two indexed lines, the RFID reader, the manipulator robot and the human inspector. Finally, to introduce production optimization into the system, a supervisor holon is also considered. The computational decisional component of these holons is implemented as software agents. The dynamic behaviour of these agents was modelled using the Petri nets formalism [10]—a suitable approach to support the formal analysis and simulation of the desired agent functionality, allowing detecting undesired behaviour or possible execution deadlocks at the design stage.

In particular, a software agent is managing the activities of the robotic device, introducing intelligence and adaptation to this automation device. The Petri net model representing the dynamic behaviour of this software agent is illustrated in Fig. 5.

The agents' behaviours were implemented using the well-known JADE framework [11], enabling the development of an intelligent and distributed architecture in a transparent manner. Despite this, and since the underground technology used in JADE is the Java™ programming language, a Java Virtual Machine (JVM) container is mandatory as the support to the developed agents. The need to have this JVM limits the number of devices that have the HW resources necessary to accommodate JADE agents. JADE agents follow the object oriented paradigm and consequently they use objects, methods and threads as the core components.

As illustrated in Petri nets model, the agent invokes several services defined in the resource interface, namely the *start* service to start the execution of a pick-and-place program, the *read* to detect the end of the robotic program and *notification* to warn about the occurrence of a failure during the program execution. The invocation of these services is made in an undistinguished manner, and without knowing the particularities of the automation device.

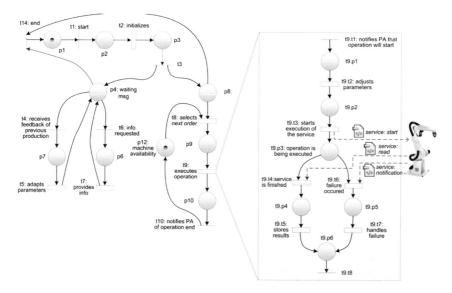

Fig. 5 Petri net model for the behaviour of the Operational Holon

4.3 Integrating the Automation HW Device

Having the interface defined, particularized instantiations of those services are performed according to the existing HW in the system (and particularly their controllers). Illustratively, two different examples, implementing the read of a bit, one using the OPC connection to a server and the other using a Modbus command, are described in the following.

```
public Boolean read (String var ) {
   ...
   JIVariant intaux = null;
   String straux = null;
   try {
     final Item item = group.addItem (var);
     item.setActive (true);
     intaux = item.read (false).getValue();
     straux=intaux.toString();
     String straux1 = straux.substring(2, 3);
   }
   catch ( final JIException e ){e.printStackTrace();}
   return(extractValue(straux1));
}
```

where the *var* parameter contains the specification of the PLC type and address extracted from the *xml* configuration file. The same read interface, using a Modbus communication protocol is now recoded using the following code excerpt.

```
public Boolean read (String var) {
    ...
  try {
     int regReference = Integer.parseInt(var);
     rcreq=new ReadCoilsRequest(regReference, 1);
     trans=new ModbusTCPTransaction(con);
     trans.setRequest(rcreq);
     trans.execute();
     rcres=(ReadCoilsResponse)trans.getResponse();
  }
  Catch(Exception e){e.printStackTrace();}
    return(extractValue(rcres.getCoils().toString().trim());
}
```

More examples could be given using the same approach, namely interfacing different robot controllers or using different communication infrastructures, e.g. a serial communication channel.

It is worthy to mention that all the aforementioned examples use a decoupled approach, where the agent control layer, due to the JVM needs, is not directly deployable into the controlled HW, i.e. into the robot controller. Despite this, the standard interfaces approach is also used when a direct HW control can be performed using an agent coupled approach, as in the case of a Raspberry Pi [9].

The experimental tests show that this approach simplifies the development and deployment of agent-based systems in the control of physical devices. On one side agents developers can only focus on developing the desired agents' functionalities, while, on the other side, automation integrators can focus on developing these interfaces, parameterized according to the particularities of the physical HW devices.

5 Conclusions

This paper presents a simple and effective approach to develop standardized interfaces that can be used to access physical automation components by the cyber layer in CPS. The proposed methodology uses the MMS standard as the ground base for the specification of the interface layer. This abstraction layer allows a fast integration of the agent with the physical world, being only necessary the implementation of the service interfaces and the parameterization of an *xml* like file that maps the used parameters, or tags, in the agent development to the real physical/logical address.

In ADACOR holonic architecture, at this stage, only the OHs use the proposed methodology. Despite this, all the other holons can benefit from this approach, where, e.g., the SH has access to different mathematical solvers, and the THs and PHs have access to their different data sources, through the use of a common interface to access different legacy systems.

In this work, this approach was exemplified to build a robotic cyber-physical production component that is deployed in a cyber-physical system for a small-scale production system. Additionally to the simplicity of the proposed approach, it turns out to be a very effective solution allowing the fast development and deployment of industrial agent-based systems.

The experimental development, parameterization and deployment of the agent based system were successfully achieved. In fact, after having the underlying OHs agents' logic developed, the tasks needed to fully complete the integration were to instantiate the libraries of the different communication protocols present at the system, namely an OPC server, a Modbus communication and a serial protocol, and to create the parameterization files. Although every communication protocol could be wrapped around the OPC server, it was opted to develop this set of libraries in order to further push this methodology.

As future work, this approach will further forester the CPS where higher level applications could use the developed services to compose more complex services, e.g., a SCADA system could use a service that is also used by the agents. Additionally, this approach must be deployed in more test beds. Only at this point one can truly take advantage of this, since, expectedly, the deployment efforts would be greatly reduced.

References

1. Leitão, P., Colombo, A.W., Karnouskos. S.: Industrial automation based on cyber-physical systems technologies: prototype implementations and challenges. In: Accepted for publication in Computers in Industry, Elsevier (2015)
2. ISO/IEC 9506-1: Industrial Automation Systems—Manufacturing Message Specification. Part 1—Service Definition (1992)
3. Leitão, P., Restivo, F., ADACOR: a holonic architecture for agile and adaptive manufacturing control. Comput. Ind. **57**(2), 121–130 (2006)
4. Schmid, M.: Cyber-Physical Systems ganz konkret, ELEKTRONIKPRAXIS, n. 7 (2014)
5. Wooldridge, M.: An Introduction to Multi-Agent Systems. John Wiley & Sons (2002)
6. Leitão, P., Karnouskos, S.: A survey on factors that impact industrial agent acceptance, In: Leitão, P., Karnouskos, S. (eds.) Industrial Agents: Emerging Applications of Software Agents in Industry, pp. 401–429. Elsevier (2015)
7. Winkler, M., Mey, M.: Holonic manufacturing systems. Eur. Prod. Eng. (1994)
8. Ribeiro, L.: The design, deployment, and assessment of industrial agent systems. In: Leitão, P., Karnouskos, S. (eds.) Industrial Agents: Emerging Applications of Software Agents in Industry, pp. 45–63. Elsevier (2015)
9. Dias, J., Barbosa, J., Leitão, P.: Deployment of industrial agents in heterogeneous automation environments. In: Proceedings of the 13th IEEE International Conference on Industrial Informatics (INDIN'15), pp. 1330–1335. Cambridge, UK 22–25 July 2015
10. Murata, T.: Petri nets: properties. Anal. Appl. IEEE **77**(4), 541–580 (1989)
11. Bellifemine, F., Caire, G., Greenwood,D.: Developing Multi-Agent Systems with JADE. Wiley (2007)

Part VII
Smart Grids and Wireless Sensor Networks

Multi-Agent Planning of Spacecraft Group for Earth Remote Sensing

Petr Skobelev, Elena Simonova, Alexey Zhilyaev and Vitaliy Travin

Abstract The paper suggests a multi-agent approach to planning of a spacecraft group for Earth remote sensing. The main constraints as well as assessment criteria of a spacecraft group planning efficiency are provided. Such a planning method is suggested, where the sought work schedule is built as a dynamic balance of spacecraft agents, observation area agents and data receiving point agents. Classes of agents as well as protocols of their interaction are described in the paper. There are considered conflict situations that occur between agents. The paper reports a planning system developed for target application of a spacecraft group, and proves the advantages of multi-agent approach to the management of a spacecraft group for Earth remote sensing.

Keywords Multi-agent technology · Spacecraft group · Earth observation · Ground station · Planning and scheduling

1 Introduction

One of the most perspective trends in the field of Earth remote sensing (ERS) is creating a multi-satellite orbit group that allows for increasing frequency of Earth surface examination as well as reliability and viability level of the space system. Expansion of an orbit group results in alternative possibilities of observing the same areas with various spacecraft. At the same time, with the limited number of data receiving points (DRPs), it becomes inevitable when several spacecraft lay claim to

P. Skobelev · E. Simonova · A. Zhilyaev (✉)
Samara State Aerospace University, Samara, Russia
e-mail: zhilyaev@smartsolutions-123.ru

E. Simonova
e-mail: simonova@smartsolutions-123.ru

E. Simonova · A. Zhilyaev · V. Travin
SEC Smart Solutions Ltd., Samara, Russia
e-mail: travin@smartsolutions-123.ru

© Springer International Publishing Switzerland 2016
T. Borangiu et al. (eds.), *Service Orientation in Holonic and Multi-Agent Manufacturing*, Studies in Computational Intelligence 640,
DOI 10.1007/978-3-319-30337-6_28

transmitting data to the same DRP. Increase of interest to Earth remote sensing results in the need for allocation of observation requests considering their priorities and execution time. Hence, the necessity of real-time dynamic coordination of a group resource functioning plan occurs.

However, most of the existing developments in this area are aimed at creating static resource usage plan of the space system [1]. At that, there is a disputable assumption that a spacecraft functions in a determined environment. Besides, the used planning methods and means are primarily aimed at separate spacecraft and cannot be projected to large-scale groups.

With the introduction of multi-satellite groups and the corresponding growth of their target functioning complexity, various heuristic algorithms have been suggested for solving this task. This question has been widely discussed in [2]: authors compare several implementations of genetic algorithms that are combined with hill climbing, simulated annealing, squeaky wheel optimization and iterated sampling algorithms. The research in [3] considers planning of ERS tasks that occur continuously and asynchronously by means of ant colony optimization algorithm. According to [4], for solving the task of spacecraft planning a combination of artificial neural network and ant colony optimization algorithm is suggested. Dynamic planning of spacecraft operations is separately considered in [5]. The literature shows that the principle of adaptive spacecraft resource scheduling with the use of heuristic methods is very efficient. The paper considers the possibility of implementing this principle with the help of multi-agent technology, which showed good results when solving traditional tasks of resource planning and allocation [6].

2   Problem Statement

Let us assume that at the initial time a set of $j = \overline{1,m}$ spacecrafts, $k = \overline{1,k}$ DRPs and $i = \overline{1,n}$ observation areas is given, as well as the duration of equipment operation during shooting τ_{shoot} and data transmission τ_{drop}. Visibility cyclograms of spacecraft and DRPs are given as well as time periods in which spacecraft are staying within certain observation areas. One has to plan shooting of observation areas by spacecraft groups, which have certain memory productivity and capacity level, with the subsequent transmission of the images to the network of DRPs. Minimizing storage time of shots in the spacecraft on-board memory unit (1) is an important criterion of efficiency of the developed schedule.

$$L = \frac{1}{n}\sum_{i}^{n}\left(1 - \frac{t^i_{shoot} - t^i_{drop}}{t_{max}}\right) \rightarrow max, \quad where: \tag{1}$$

t^i_{shoot}—time when shooting was started for the observation area i by a chosen spacecraft;

t_{drop}^i—time when the shot transmission was started for the observation area i to a chosen DRP;

t_{max}—critical storage time for a shot, after which it is considered outdated.

The developed schedule must satisfy the following constraints:

1. Visibility between a spacecraft and an observation area during shooting.
2. Visibility (accessibility) between a spacecraft and a DRP during data transmission.
3. Free space in the on-board memory unit of a spacecraft (2).

$$\sum_i^n q_j \varphi_{ij}(t) \leq Q_j, \text{ for } j = \overline{1,m}, \text{ where} : \tag{2}$$

q_j—volume of information contained in one shot of a spacecraft j;

$$\varphi_{ij}(t) = \begin{cases} 1, & \textit{shooting of area i is done by a spacecraft j at the moment t,} \\ 0, & \textit{in other cases}; \end{cases}$$

Q_j—storage capacity of an on-board memory unit of spacecraft j.

4. Concurrency of the times for shooting, data transmission and receiving (3–4).

$$t_{shoot}^i + \tau_{shoot} < t_{drop}^i, \text{ for } i = \overline{1,n}, \tag{3}$$

$$t_{drop}^i = t_{receive}^i, \text{ for } i = \overline{1,n}, \text{ where} : \tag{4}$$

$t_{receive}^i$—time when receiving the shot was started for the observation area i by a chosen DRP.

5. No overlapping between operations in schedules of different resources (satellites and DRPs are forbidden to perform several operations simultaneously) (5–6).

$$\sum_i^n \varphi_{ij}(t) \leq 1, \text{ for } j = \overline{1,m}, \tag{5}$$

$$\sum_i^n \omega_{ik}(t) \leq 1, \text{ for } k = \overline{1,k} \text{ where,} \tag{6}$$

$$\omega_{ik}(t) = \begin{cases} 1, & \textit{shot receiving of the observation area i by a DRP k at the moment t,} \\ 0, & \textit{in other cases}. \end{cases}$$

An important feature of the task is taking into account dynamically occurring events, including introduction of a new task or change of task options, failure of spacecraft resource or means of communication, inaccuracy or error of achieving shooting results, etc.

3 Methods

3.1 Description of the Approach

In comparison with the traditional technology that presupposes static planning of shooting and data transmission sessions, the paper suggests using multi-agent approach where the sought functioning plan of a spacecraft group is built by self-organization of separate agents based on their competition and cooperation. A concept of demand-resource networks is applied, where each schedule is designed as a flexible (rebuilt by the events) network of connected demand (problem) and resource agents. It is based on the principle of joint interest of all the participants in the solutions profitable for each of them and for the system as a whole. At the same time, worsening of one participant's position can be compensated at other participants' expense in the interest of the group if this eventually leads to the benefit of the group as a whole.

Interaction of agents and change of the corresponding orders and resources results in an acceptable locally optimal decision, which is adaptively adjusted in "a sliding mode" at the considered planning horizon. At that, a possibility of adaptive change of the previously made plan is added (without its complete rearrangement). This means that the plan is not built every time when new events occur (as in traditional methods of operation research) but is only corrected without system restart [7, 8].

In the context of planning Earth remote sensing task, the sought schedule is built as dynamic balancing interests of spacecraft, DRP and observation area resource agents. Besides, a system agent is introduced which is capable of assessing performance of other agents and assigning strategies most suitable for them in a certain situation. An observation area agent has demands for operations of shooting, as well as transmission, storage and receiving the information about the shot. These demands can be satisfied with the help of resource agents—spacecraft and DRPs.

Efficiency criteria of the formed schedule are reflected in target functions of agents [9]. The observation area agent strives for shooting of its area and transmission of data to a DRP as quickly as possible, with minimal storage period in an on-board memory unit of a spacecraft. The aim of resource agents of spacecraft and DRPs is to increase productivity at the considered planning horizon. The target function of the system is determined by the total sum of target functions of separate agents.

All the constraints mentioned in problem statement are distributed among the corresponding agents. For example, when planning mutual visibility cyclograms of spacecraft, observation areas and DRPs are considered as well as their parameters. Each image of observation area has certain capacity and fills up the on-board memory unit. It results in the necessity of data transmission, otherwise a spacecraft stops shooting other areas.

Formation of the sought schedule of shooting and data transmission between spacecraft and DRPs is divided into two stages: *conflict-free* and *proactive*.

3.2   Initial Conflict-Free Planning

At the first stage agents of observation areas send requests to suitable spacecraft about the possibilities of shooting and transmitting to the Earth within the best (according to the target function) free time interval. The decision obtained at this stage is taken as the initial one, which will further be consistently improved starting with the "worst" fragments of the plan. Spacecraft agents have access only to the timetables of their own spacecraft's resources. According to this data, they make decisions about the possibility or impossibility of placing a new image of the observation area. If at the time of receiving a request from the observation area all resources of the spacecraft are already occupied, its agent declines the request for shooting of the area.

Conflict-free route planning is possible if all of the following conditions are fulfilled:

- Visibility between a spacecraft and an observation area and a DRP;
- The time period does not overlap with the previously planned data transmission session or shooting of other observation areas;
- Spacecraft on-board memory unit contains sufficient amount of free space for the time interval from the beginning of shooting of the observation area to the end of image transmission to DRP.

The spacecraft agent sends inquiries to all the known DRP agents with a proposal to hold a data transmission session. Among all the options proposed by DRP agents, the spacecraft agent chooses the closest to the moment of shooting time interval, but at the same time the one that is free from other shooting sessions or DRP communication sessions. If any of the conditions is not fulfilled, the shooting of the observation area remains unscheduled. Having planned a DRP communication session, the spacecraft agent informs the observation area agent about the data transmission time and makes the necessary changes in its timetable. It is important to note that the purpose of this stage is to quickly obtain a feasible initial schedule, whatever its level of quality. The solution received at this stage shows the main bottlenecks of the timetable and becomes the reference point for further improvements.

3.3   Proactive Schedule Improvement

At this stage, the observation area agents are trying to improve the value of their target function, suggesting that conflicting with their areas needs finding other intervals for placement by shifting the time or moving to another resource (spacecraft or DRP). Building a sequence of changes is started by those agents that are most unsatisfied with the value of their target function. A proactive observation area agent asks available resources about the possibility of placing certain operations; then, some conflicts are inevitably exposed: time slots that are favourable from the point of view of the target function are found to be occupied by other operations. Those agents that are connected with these operations receive a request for a shift to a specified time slot. Recursive shifting of the operations affected by the shift continues until one of the operations can move to a new position without any obstacles; the displacing operation proceeds as long as there are means to compensate the induced expenses or until a counter which limits recursion depth equals zero. Such a process of agent interaction when shifting the operations in the schedule is shown in Fig. 1: operations are symbolized by rectangles of varying width proportional to their duration; the displacing operation is shaded, solid arrows represent messages generated by the shift request, and response messages of the shifted operations are shown as dotted lines.

The following conflict situations are taken into consideration when building a chain of changes:

1. Planning of shooting in the observation area by displacement of the previously planned shooting sessions or data transmission sessions from the spacecraft schedule.
2. Approximation of the time of image transfer to DRP by displacement of the previously planned shooting sessions or data transmission sessions from the spacecraft schedule.
3. Displacement of the previously planned data transmission sessions from the DRP schedule.
4. Emptying the spacecraft memory unit of other images in case of lacking space in the on-board memory unit.

Fig. 1 Recursive operation shifting

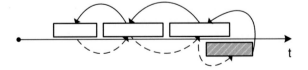

Efficiency of each permutation can be evaluated by changing the values of the target functions of participating agents. The resulting version of the schedule change can be either accepted or rejected depending on the effect exerted on the target function of the system. Only those changes are approved which do not reduce the target function of the system. The task that remains unscheduled is placed into the list of tasks that wait for being scheduled. A new attempt of scheduling these operations will be made in case of arising events such as adding new resources or changes in schedules of the existing ones. Figure 2 shows the negotiation log of agents at the proactivity stage.

Similarly, agents of spacecraft resources and DRPs are trying to increase the target function value for the system through reallocation of shooting and data transmission sessions. Iterative improvement of the schedule by all types of agents continues until the state of "dynamic stop" is reached (agents are trying to improve their condition but cannot achieve improvement of the target function value). This will mean reaching a consensus in the negotiations and the possibility of issuing the final solution. This state can be disturbed if certain events occur which are connected to adding, deleting or changing the parameters of spacecraft, DRPs or observation areas. In this case the schedule is reduced to an acceptable form and then it is improved via the above-described mechanism.

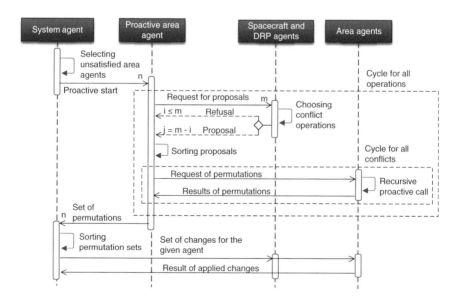

Fig. 2 Negotiation log of agents at the proactivity stage

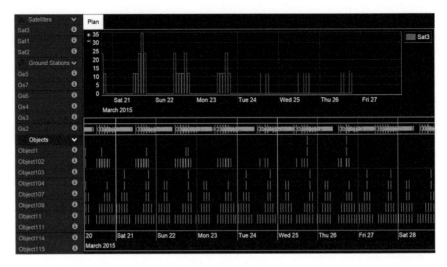

Fig. 3 A screen of multi-agent prototype system for planning of a spacecraft group for ERS

4 Results

The method described above has been implemented in a prototype multi-agent system for planning the use of spacecraft for Earth remote sensing. The system has a client-server architecture. The input data required for planning is calculated by means of third-party software that takes into account the ballistic model of spacecraft movement and the geographical location of observation areas and DRPs. The planning subsystem is implemented by means of Java Akka Framework. Input data and planning results are displayed via the user web-interface (Fig. 3). A list of spacecraft, DRPs and observation areas involved in planning is shown in the left part of the web- page. When one of the items in this list is selected, its related parameters and schedule are shown in the right part of the page. When a certain spacecraft is selected, a diagram of its on-board memory unit fullness is displayed at the top of the page. Any change in the input data automatically leads to reconstruction of the existing schedule.

5 Conclusion

Within the proposed approach, the problem of space system management is solved by creating a self-organizing team of intelligent agents conducting negotiations and not only capable of planning their behaviour individually in real time, but also working in groups in order to ensure coordination of decisions.

The developed software prototype for planning the use of spacecraft for Earth remote sensing has proved the potential of this approach owing to the following facts:

- Significant reduction of time spent on forming a schedule close to optimal (compared to the exhaustive algorithm);
- Flexibility provided by a rapid response to emerging events;
- Scalability and openness—new components (spacecraft, DRPs, etc.) can be connected to the system dynamically without the system's shutdown and restart;
- Autonomy of software modules which will at long term make it possible to place planning components inside the spacecraft's on-board computing devices [10].

Acknowledgements This work was carried out in SEC "Smart Solutions" Ltd. with the financial support of the Ministry of Education and Science of the Russian Federation (Contract № 14.576.21.0012, unique number RFMEFI57614X0012, and P. Skobelev—scientific advisor, E. Simonova—senior analyst, V. Travin—project manager, A. Zhilyaev—programmer).

References

1. Sollogub, A., Anshakov G., Danilov V.: Spacecraft systems for sensing of the Earth's surface. Mechanical Engineering, Moscow (2009)
2. Globus A., Crawford J., Lohn J., Pryor A.: Application of techniques for scheduling earth-observing satellites. In: Proceedings of the 16th conference on Innovative Applications of Artificial Intelligence, 836–843 (2004)
3. Iacopino, C., Palmer, P., Policella, N., Donati, A., Brewer, A.: How ants can manage your satellites. Acta Futura **9**, 57–70 (2014)
4. Rixin, L.Y.W., Xu, M.: Rescheduling of observing spacecraft using fuzzy neural network and ant colony algorithm. Chin. J. Aeronaut. **27**, 678–687 (2014)
5. Chuan, H., Liu, J., Manhao, M.: A dynamic scheduling method of earth-observing satellites by employing rolling horizon strategy. Sci. World J. (2013)
6. Rzevski, G., Skobelev, P.: Managing complexity. WIT Press, London-Boston (2014)
7. Wooldridge, M.: An introduction to multiagent systems, 2nd edn. Wiley, London (2009)
8. Skobelev, P.: Multi-agent systems for real time resource allocation, scheduling, optimization and controlling: industrial application. In: 10th International Conference on Industrial Applications of Holonic and Multi-Agent Systems, Toulouse, France (2011)
9. Belokonov, I., Skobelev, P., Simonova, E., Travin, V., Zhilyaev, A.: Multiagent planning of the network traffic between nano satellites and ground stations. Procedia Eng.: Sci. Technol. Exp. Autom. Space Veh. Small Satell. **104**, 118–130 (2015)
10. Sollogub, A., Skobelev, P., Simonova, E., Tzarev, A., Stepanov, M., Zhilyaev, A.: Intelligent system for distributed problem solving in cluster of small satellites for earth remote sensing. Inf. Control Syst. **1**(62), 16–26 (2013)

Methodology and Framework for Development of Smart Grid Control

Gheorghe Florea, Radu Dobrescu, Oana Chenaru, Mircea Eremia
and Lucian Toma

Abstract To better serve customers, operators need to identify ways to improve the reliability of their electrical service. This paper provides a methodology and framework for the development of new control architectures based on uncertainty management and self-reconfigurability. We designed an architecture for Smart Grid able to integrate the standard control strategy with safety and security aspects. A case study presents details on how the current methods for the assessment of the system security can be applied on the proposed architecture.

Keywords Smart grid · Control hub · Uncertainty management · Reconfigurable control

1 Power Grid State of the Art

Reliable energy infrastructure is a critical objective of modern society, so that any attempt to its operation has an enormous impact on people and economy.

The transmission grid uses backup lines and multiple routes for power flow to handle contingencies. An alternate route can usually accommodate the loss of a

G. Florea (✉) · O. Chenaru
Societatea de Inginerie Sisteme SIS S.A., Bucharest, Romania
e-mail: gelu.florea@sis.ro

O. Chenaru
e-mail: oana.chenaru@sis.ro

R. Dobrescu · M. Eremia · L. Toma
University Politehnica of Bucharest, Bucharest, Romania
e-mail: rd_dobrescu@yahoo.com

M. Eremia
e-mail: eremia1@yahoo.com

L. Toma
e-mail: lucian_toma_ro@yahoo.com

© Springer International Publishing Switzerland 2016
T. Borangiu et al. (eds.), *Service Orientation in Holonic and Multi-Agent Manufacturing*, Studies in Computational Intelligence 640,
DOI 10.1007/978-3-319-30337-6_29

transmission line but many parts of the distribution system do not have this capability.

The renewable energy sources are the biggest challenges for power system operators. The intermittent generation may have negative impact on the power flows, voltages as well as other network parameters. Consecutive outages may occur in a power system and thus it is highly recommended that the power system is designed so that to withstand double disconnections. Unfortunately, due to the large investment requirements, most of the power systems may have problems to comply with this criterion.

The operating conditions vary continuously and the power system moves from one state to another, as suggestively indicated in Fig. 1. N and N-1 represent standard analysis of power system operating regimes (N means all power lines are up, N-1 means one power line is down, N-1-1 means two lines are down). Transition to one state or another depends on the random events that may occur or on the decision taken by the system operator. The figure shows also a classification of possible states of the power system depending on the events that may occur.

When the power system enters the alert state, immediate corrective actions must be taken in order to restore the normal operation. If during this transition a contingency occurs, the system can enter in an emergency state, in which there are a large number of bus voltage limits violations. In this state, ultimate (extreme) actions can still be taken to restore the system to a normal operation state. If the contingency is too severe, the power system may become instable and finally collapses (as shown in Fig. 1).

Following the major incidents that have occurred in the European interconnected power system [1, 2], UCTE has issued in 2009 a new policy for operation security [3]. Maintaining the power system in secure operating state assumes that some

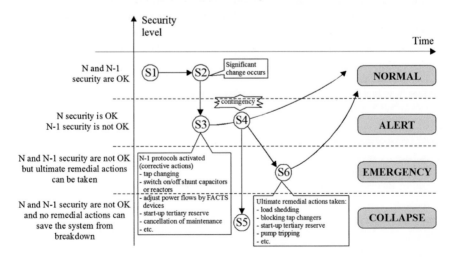

Fig. 1 Power system states [3]

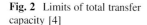

Fig. 2 Limits of total transfer capacity [4]

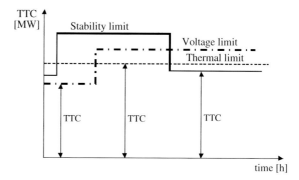

electrical quantities should be maintained within admissible limits. The most important limits that could restrict the power transfer in a transmission grid are: thermal limit, voltage limit and stability limit.

The transmission capacity of a power grid can change in time due to changes of one of the above-defined limits (Fig. 2). The total transfer capacity is given by the most restrictive limit, that is:

$$TTC = Minimum\{thermal\ limit,\ voltage\ limit,\ stability\ limit\} \qquad (1)$$

including also the case of the most severe contingency.

Exceeding one of the security limits denotes a network congestion that may jeopardize the power system operation. However, the congestions related problems are managed by technical or economical mechanisms.

In real-time, network congestions may emerge as a consequence of load forecast errors and thus, of unscheduled power flows, as well as of technical reasons such as unscheduled disconnection of a generator, an electrical line or a transformer [5]. These blackouts are the reason both for the industry and academia to consider new techniques and system architecture designs that can help assure the security and reliability of the power grid. High reliability can be designed into the distribution system—and the Smart Grid can incorporate this attribute.

A smart distribution grid emerges from decentralized control when the switching points between circuits, as well as several points along each circuit, carry the intelligence to reconfigure the circuits automatically when an outage occurs and quickly to reroute power to as many customers. More intelligent switching points yield more options to reroute power to serve the load, and communication between those points makes self-healing a practical reality.

1.1 Control Reconfiguration

The control strategy applied to processes still needs to be adequately addressed because the target is oriented not only to parameter control but more on improving overall process operability by minimizing downtimes. One of the first proposals for an architecture based on reconfigurability design is that of Choksi and McFarlane [6] which uses the coordinating function to monitor and control local planning, local optimization and local control.

Most available control theories assume that a control structure is given at the outset. There are two main approaches to the problem, a mathematically oriented approach (control structure design) and a process oriented approach [7, 8]. Plantwide control is a holistic approach concerned with the structural and functional decisions involved in the control system design of a process.

1.2 Design Power Grid Control for Uncertainties Management

Managing uncertainties in industrial systems is a daily challenge to ensure improved design, robust operation, desirable performance and responsive risk control. Uncertainty management is concerned with the quantification of uncertainties in the presence of data, model(s) and knowledge about the system, and offers a technical contribution to decision-making processes whilst acknowledging industrial constraints.

Designing for uncertainty management leads engineers to create architectures that do not meet fixed specifications. Dealing with systems attributes such as flexibility or robustness lead designers to different solutions than those that focus on optimization.

An innovative contribution is proposed in [7] as an architectural oriented approach used to develop advanced control integrated architectures, open to incorporate more flexibility, to implement reconfigurable control, to ensure the reaction to uncertainties, as well as to define a solution for integration of safety control and prevention in control strategies. RH Control is an efficient solution to control uncertainties, not only for process control but also for transmission network.

1.3 Transmission Network Reconfiguration

Reconfiguration of transmission networks is necessary for maintaining power supply during maintenance or restoration after an outage before the fault is repaired. Fully automatic reconfiguration is one of the most important attributes of the Smart Grid with a big contribution to self-healing, sustainability and flexibility [9].

Botea [10] addresses the reconfiguration problem for outage recovery, where the cost of the switching actions dominates the overall cost. Finding optimal feeder configurations under most optimality criteria is a difficult optimization problem.

2   Development of Integrated Control, Safety and Security System for Power Grid

2.1   Methodology

In order to solve the uncertainties management problem the designer will try to split it into manageable parts. A generic methodology to perform requirements analysis while addressing hazard and risk includes the following steps:

- Applicability specification (program, project, data, constraints, personnel).
- Hazards identification (expert opinion/lessons learned/test data/technical analysis/hazard analysis).
- Consequences evaluation (impact/severity, probability, timeframe).
- Risk assessment (resources/leads/project; metrics information, risk management structure).
- Monitor risk metrics and verify/validate mitigation actions.
- Checking real scenarios and progressive updating the input information.

Our approach to develop advanced control using an integrated architecture that is open to incorporate more flexibility implements reconfigurable control to ensure the reaction to uncertainties. Integration of Risk and Hazard (RH) Control able to maintain the process in a safe state using control hierarchy layers is shown in Fig. 3 where (a), (b), (c) and (d) suggest a holonic organization at several levels.

The four main holons are, from the upper level: (d)—*classical control* based on basic regulatory, sequential and logical, (c)—*safety level* based on safety instrumented systems (SIS) and the new paradigm Reconfigurable control, (b)—*remote level* based on internet or cloud able to do automatic identification, modelling and simulations, (a)—*management level*, which has two main functions: management and supervisory control [7].

To be able to perform such tasks the system architecture, structure and data flows must be able to support different methods of reconfiguration. Consequently, reconfigurability design must focus on: (i) **Components** (sensors, actuators, IEDs, synchrophasors, FACTS, controllers, equipment); (ii) **Control** (algorithms, structure, data flows, RH control strategies, integrated control); (iii) **Transmission** process (equipment, flows, process, and states).

2.2   Reconfigurable Control Architecture—Design and Applications

To develop a framework for control system design we propose the architecture presented in Fig. 4. The framework is able not only to host the entire process control system (hardware, software, application, operator and engineering interfaces) but also the model, even simplified, of the process, the simulation features, a

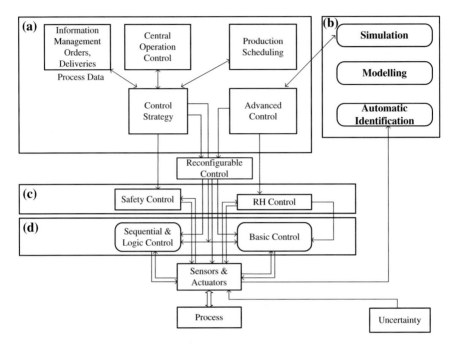

Fig. 3 Control and safety architecture adapted from [7]

library of algorithms and strategies, case studies. The focus in our work is on developing a structure of fault detection and intelligent alert that in conjunction with *Reconfigurable Control* can conduct to the recovery of functionality, even with spoiled performances. Mode selection—part of this structure functions as follows: at first, the fault recovery measures for individual loop failures are derived from a fault impact analysis. Next, the fault recovery principle initiates a change in the operating strategy of the plant by incorporating changes in the operating factors associated with failures in the model based control calculations. These strategies can be implemented with direct commands from Reconfigurable Control or/and associated with reconfiguration scenarios.

3 Case Study

The Romanian power system has undergone significant changes concerning the generation pattern. In December 2014 the total installed capacity in wind power plants (WPP) was 2950 MW. At least 80 % of the capacity installed in WPPs is located in the Dobrogea region. A 1400 MW rated power nuclear power plant is also connected in this region and operates at full capacity. The average load in the

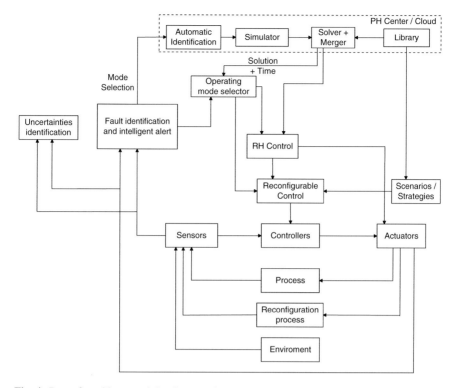

Fig. 4 Reconfigurable control development framework

Romanian power system is about 6700 MW, while an average power of 800 MW is exported to other power systems.

The network section (Fig. 5) is defined across the electrical lines that interconnect the Dobrogea region with other parts of the Romanian power system or with the Bulgarian power system. These lines are also the most subjected to overloading.

Fig. 5 A simplified representation of the Dobrogea region

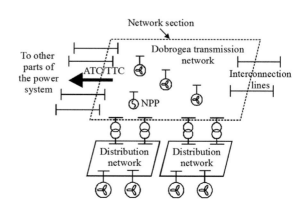

From the security point of view several limits are verified both in the planning activity one day before, and also in exploitation which takes place in real-time.

The thermal limit is verified one day before for both N and N-1 configurations; if overloading is identified, network reconfiguration or/and generation dispatching is performed. In real-time if N-1-1 contingency occurs, either immediate action by the dispatcher is taken or appropriate automation is activated by reducing the power generation within the area. For this purpose, a study was performed to define a regional automation scheme that deals with any unexpected contingency. Taking into account the large number of scenarios, the influences (sensitivities) of all wind power plants on the power flows on the transmission lines have been determined. A ranking of these sensitivities have been roughly defined so that, in real-time, when a certain contingency occurs, the minimum quantity of generated power is disconnected.

The bus voltages are regulated in two stages to be maintained within predefined voltage limits. The first stage consists in providing the voltage set-points to the nuclear power plant and the nodes where large wind power plants are connected. The second stage is the real-time operation and consists in voltage control contribution to the pilot buses from the wind power plants. However, this stage is active if there is generation availability from wind power plants. When active, voltage set-points roughly about 105 % of the nominal value are set within area because Dobrogea region becomes an important reactive power sources for other parts of the Romanian power system.

The stability limit is calculated one day before by off-line simulations and in real-time (on-line) by specialized software. The stability limit is defined as the maximum power that can be transited through the predefined section. Since the Dobrogea region exports power, the stability is the sum of power flows on all lines from the section. This limit decreases when one or more lines are disconnected; in this case the most severe contingency is identified, which gives the stability limit for the N-1 configuration, a.o. The stability limit is calculated one day before and if the scheduled exported power exceeds these limits some actions are taken. The same study had the purpose of defining an automation logic that takes actions in real time when the stability criterion is not met. The automation logic aims to reduce the power produced by wind power plants when the stability limit is exceeded, mainly in case of contingencies.

4 Conclusions

Our work promotes the concept of holonic control based on uncertainties management instead of using the standard control strategy approach. Real-time capability jointly integrated with Smart Grid attributes like isolation, reconfiguration, modularity and standardization provide the necessary tools for uncertainties management and lead to a more reliable system. This approach was considered in simulating the behavior of the power system in Dobrogea region. Future work will include identifying a particular solution for risk and hazard control with self-reconfiguration of a pilot plant.

Acknowledgements This work was supported by the Romanian National Research Programme PNII, project: Intelligent buildings with very low energy consumption—SMARTBUILD.

References

1. Berizzi, A.: The Italian 2003 blackout. In: IEEE-PES General Meeting, vol. 2, pp. 1673–1679, IEEE, Denver (2004)
2. Final report on system disturbance on 4 November 2006. Technical report, Union for the Coordination of Transmission of Electricity (2007)
3. UCTE Policy 3: Operational security. Technical report, Union for the Coordination of Transmission of Electricity (2009)
4. Available transfer capability definitions and determination: a framework for determining available transfer capabilities of the interconnected transmission networks for a commercially viable electricity market. NAERC (1996)
5. Eremia, M., Shahidehpour, M., et al. (eds.): Handbook of Electrical Power System Dynamics: Modeling, Stability, and Control. Wiley & IEEE Press, Power Engineering Series, Hoboken (2013)
6. Chokshi, N.N., McFarlane, C.D.: A Distributed Coordination Approach to Reconfigurable Process Control. Springer Series in Advanced Manufacturing (2008)
7. Dobrescu, R., Florea, G.: Integrating risk and hazard and plantwide control solutions for reconfigurability. In: Borangiu, T., Thomas, A., Trentesaux, D. (eds.) Studies in Computational Intelligence. Service Orientation in Holonic and Multi-agent Manufacturing, vol. 594, pp. 103–114. Springer (2015)
8. Pournaras. E., Yao, M., Ambrosio, R., Warnier, M.: Organizational control reconfigurations for robust smart power grid. In: Besis, N., Xhafa, F., Varvarigou, D., Hill, R., Li, M. (eds.) Studies in Computational Intelligence. IoT and Inter-cooperative Computational Technologies for Collective Intelligence, vol. 460, pp. 189–206. Springer (2013)
9. Kok, K., Scheepers, M., Kamphuis, R.: Intelligence in electricity networks for embedding renewables and distributed generation. In: Negenborn, R.R., Lukszo, Z., Hellendoorn, H. (eds.) Intelligent Systems, Control and Automation: Science and Engineering, Intelligent Infrastructures, pp. 179–209. Springer (2010)
10. Botea, A., Rintanen, J., Banerjee D.: Optimal reconfiguration for supply restoration with informed A* search. In: IEEE Transactions on Smart GRID, vol. 3, no. 2, pp. 583–593. IEEE (2012)

Sink Node Embedded, Multi-agent Systems Based Cluster Management in Industrial Wireless Sensor Networks

Mohammed S. Taboun and Robert W. Brennan

Abstract With advances in cyber-physical systems and the introduction of industry 4.0, there has been an extensive amount of research in distributed intelligent control. Because in cyber-physical systems it is required that devices are aware of their environment, industrial wireless sensor networks are considered for such types of applications. In this paper, a sink node-embedded multi-agent system is proposed in order to manage clusters of wireless sensors; this architecture is analysed in an oil and gas refinery example.

Keywords Multi-agent systems · Wireless sensor networks · Distributed embedded intelligence

1 Introduction

With advances in cyber-physical systems and the introduction of Industry 4.0, there has been an extensive amount of research in distributed intelligent control. The requirement of cyber-physical system is that devices be aware of their environment; industrial wireless sensor networks (WSNs) have been considered for this application. Research has shown that multi-agent systems have proven to be a successful technique for managing WSNs. Typically this is done through a coupled or cloud based deployment. With advances in technology it is becoming feasible to deploy these intelligent agents directly on the automation hardware.

Key challenges in WSNs are fault recovery and scalability, especially in industrial systems situated in harsh environments. WSNs' scalability depends tremendously on the ease of introducing new sensor nodes into the network.

M.S. Taboun (✉) · R.W. Brennan
University of Calgary, Calgary, Canada
e-mail: mstaboun@ucalgary.ca

R.W. Brennan
e-mail: rbrennan@ucalgary.ca

© Springer International Publishing Switzerland 2016
T. Borangiu et al. (eds.), *Service Orientation in Holonic and Multi-Agent Manufacturing*, Studies in Computational Intelligence 640,
DOI 10.1007/978-3-319-30337-6_30

329

This ease-of-introduction characteristic is also one of the key challenges in Industry 4.0 systems.

In this paper, an embedded multi-agent system for managing sink nodes and clusters of wireless sensor networks is proposed, and finally is demonstrated in an oil refinery application. The network architecture can be used to develop protocols for easily introducing new sensor nodes into the WSN, as required by WSN. The agent platform developed can be used for future tests on fault recovery.

2 Background

In this section, background information is given on cyber-physical systems, wireless sensor networks and industrial agents.

2.1 Industry 4.0 and Cyber Physical Systems

With advances in technology, cyber-physical devices are beginning to emerge in industrial applications. A cyber-physical system is a set of cyber-physical devices that according to reference [1] include computing that controls mechanical activity through embedded processing, networking and connectivity, awareness of the environment and other objects through sensors, and finally a means of interacting with the environment through actuators.

A new trend in research which employs these cyber-physical systems is known as Industry 4.0. According to reference [2], the term Industry 4.0 describes different changes in manufacturing systems with not only technological but organizational implications. These changes are expected to shift from product to service orientation in industrial systems. This can be extended from manufacturing systems to other types of industrial systems, such as oil and gas refineries. These shifts will lead to new types of enterprises which adopt new specific roles within industry.

2.2 Industrial Wireless Sensor Networks

As previously mentioned, cyber-physical systems require awareness of the environment and other objects through sensors. Industrial wireless sensor networks (WSNs) have become a research trend due to the advances in processing power for micro-computers and reduced battery consumption of embedded battery powered devices. WSNs are composed of wireless sensor nodes, which are small with limited processing and computing resources and are inexpensive compared to traditional sensors [3]. Sensor nodes are used to sense, measure and gather information from the environment and transmit the data to a user or data acquisition system.

One of the primary concerns in wireless sensor network is data routing. When a large scale industrial WSN passes a lot of data, this creates a large communication overhead. The most widely accepted solution to reduce this overhead is to cluster the wireless sensor network. This clustering process forms a hierarchy structure for the network and allows for data aggregation. This hierarchy can then be composed of two types of sensor nodes: *sink nodes* and *anchor nodes*.

Sink nodes are the cluster-heads of the network. They are responsible for aggregating the data and transmitting information from the network to the acquisition system or base station. Due to the fact that there are many transmissions and data aggregation required of the sink node, it is often a higher processing, fixed unit. The sensor nodes which make up the cluster and send sensory data to the sink nodes are often referred to as anchor nodes.

In some WSNs, the sink node is a regular wireless sensor node. This is referred to as a homogeneous WSN (as opposed to a heterogeneous WSN). In this case, the extra processing power required of the sink node drains the battery at a faster rate than the anchor nodes, and sink node rotation becomes a primary concern to prolong the network lifetime. In industrial applications that require a perpetual lifetime of the wireless sensor network, it is often more practical to use a heterogeneous wireless sensor network. This reduces the complexity of sensor node replacement and maintenance programs.

2.3 Industrial Agents and Agent Based Control

While there are several definitions of agents, the most commonly accepted definition is provided by reference [4], which states that an agent is a computer system that is situated in some environment, and that is capable of autonomous action in this environment in order to meet its delegated objectives. Agents are also often defined by their characteristics. According to reference [5], agents are autonomous, responsive, proactive, goal-oriented, smart-behaving, social and able to learn.

A multi-agent system is a system of two or more agents that collaborate to some sort of collective goal, while still working to their own individual goals. According to reference [5], multi-agent systems have decentralized control and are flexible, adaptable, reconfigurable, scalable, lean and robust. The properties of multi-agent systems align with the design considerations for wireless sensor networks, and are therefore well suited to manage these networks.

A major point of interest with the advances in technology is whether to embed the intelligent agent or to use a coupled or cloud based design. Reference [5] defines a coupled design as a situation where one or more agents collect and process data from an existing structure, in a cloud-based fashion. Embedded agents are when the automation platform itself is agent-based. While the coupled design can be immediately applicable and integrate with existing technology, advances in controllers are allowing the embedded intelligent design to become feasible.

3 Related Work

3.1 Agent-Based Wireless Sensor Networks

As previously mentioned, the distributed nature of multi-agent systems aligns with the distributed properties of wireless sensor networks. For this reason, many researchers have proposed using adapting, intelligent agents to work on distributed and complex sensor networks. In this section, we describe some of these works.

In Ref. [6], multi-agent solutions for WSNs are examined. A multi agent architecture which interconnects a wide range of heterogeneous devices that may possess various levels of resources is proposed in Ref. [7]. Similarly, Ref. [8] proposed the development of intelligent sensor networks using multi-agent systems. In this work, the multi-agent system was implemented in the Java Agent Development framework (JADE). Ref. [9] compared alternative cluster management approaches using multi-agent systems. This set of simulations also saw the multi-agent system implemented in JADE.

A multi-agent based application oriented middleware is introduced by Ningxu et al. [10], in which a multi-agent management system controls a distributed control system via IEC 61499 function block controllers. The middleware of a WSN refers to a set of tools that reduce the complexity on lower level hardware systems. In traditional PLC based systems, this approach is often examined in order to build intelligence into simple programmable controllers. This middleware is designed specifically for wireless sensor networks that track mobile objects in factory automation.

3.2 Wireless Sensor Networks for Oil Refineries

Research in multi-agent systems and wireless sensor networks is typically concerned with manufacturing applications. There are however some research areas that bring this technology to other industries, such as health care or safety. One of these industries is oil and gas refineries.

In Ref. [11], the most promising wireless technologies were examined in order to cope with the challenges in implementing a WSN in an oil and gas refinery. Gil et al. [12] proposed an outlier detection and accommodation methodology for oil refineries using wireless sensor networks. The model was tested on a real monitoring scenario implemented in a major refinery plant.

4   Agent Based, Sink-Node Embedded Intelligence Model

In this section, an embedded multi-agent system for managing clusters is proposed. The first sub-section outlines the individual agents including their knowledge, skills and communication and is followed by how the agents interact with each other.

4.1   Agent-Based Sink Node Management

Each sink node has three agents that provide the intelligence required to manage its respective cluster. These agents are the *sink node mediator*, *device manager* and *task manager*, and are shown in Fig. 1.

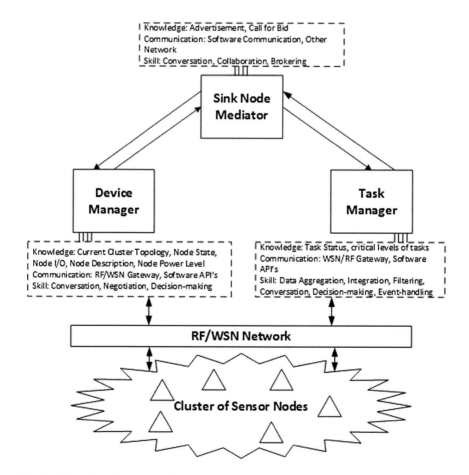

Fig. 1 Sink-node embedded multi-agent system

The device manager agent has knowledge of the current cluster topology, the hierarchic level of the sensor network, the state of the nodes in the cluster, the node I/O's, the node description and the node power levels. The device manager communicates with the nodes in the cluster via a RF gateway (such as xBee transmitters). The device manager has skills conversation, negotiation and decision making. These skills allow the device manager agent to dynamically reconfigure the nodes in the sink nodes' respective cluster.

The task manager agent has knowledge of the sensing and control task status for the sink nodes' cluster of sensors. It also has knowledge of the level of parameter being sensed and the corresponding reaction required for the control units in the cluster. Much like the device manager agent, the task manager agent communicates with the sensor nodes via RF or wireless sensor network gateways. The task management agent has a skillset of data aggregation, integration, filtering, conversation and decision-making. While the device manager and task manager agent communicate with the nodes in their cluster of sensors, they do not communicate with each other. Instead, communication between the agents is done through the sink node mediator agent. The sink node mediator has knowledge of advertisements and bidding. It communicates with the other agents on the same sink node via software communication, and to the data acquisition and other sink nodes via other networks such as LAN or Wi-Fi. The skillset of this agent is in conversation collaboration and brokering.

4.2 Wireless Sensor Network Architecture

At the highest level of the network is the supervisory control and data acquisition (SCADA) system. Due to the intelligent agents being embedded on the physical network, only simple control is required for simple inputs into the wireless sensor network. For this reason, a SCADA system is preferable over a more complex distributed control system. The SCADA system transmits and receives data from the sink nodes through a wired or wireless network.

The device manager agent is responsible for ensuring that the sensor cluster is able to perform the tasks required of it. It accomplishes this task by dynamically reconfiguring the topology of the cluster in an on-the-fly fashion. Dynamic reconfiguration considers any changes to the topology including replacing sensor nodes that have failed by either hardware/software errors or battery failure. Other types of reconfiguration include the sleep and wake function of sensors, using sensors from other cluster in the case of fault recovery and lending sensors to other clusters.

The task manager is primarily used for obtaining the application specific data from the sensor cluster. In the case of a hierarchical wireless sensor network in which a cluster is composed of sub-clusters, the task manager is aware of which level of the hierarchy it is obtaining data from. For complex sensing data, a large amount of data may be transmitted through the WSN. Data aggregation helps lower

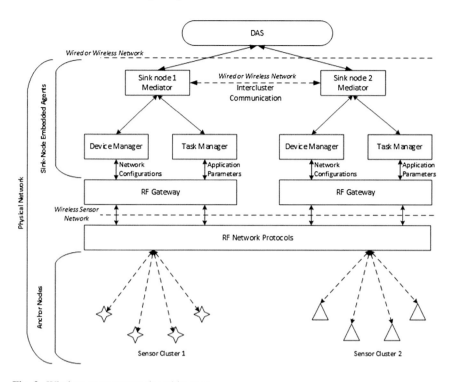

Fig. 2 Wireless sensor network architecture

the resulting overhead power consumption. The aggregated data routing can be seen in Fig. 2.

In order to ensure that the proper data is being received, some collaboration is required with the device manager. The sink node mediator is responsible for handling this collaboration. The sink node mediator is also responsible for handling inter-cluster communication, which may occur if the cluster needs to borrow another node. In this case, the device manager agent would send a request to the sink node mediator which would negotiate with the sink node mediator in another cluster.

5 Application Specific Example in an Oil and Gas Refinery

To illustrate an application of the architecture, consider the example of a simple oil refinery adapted from Ref. [13]. The simple refinery process consists of 4 major types of equipment: separators, compressors, water treatment and storage. There are two types of separators: stage 1 and stage 2. There are also two types of gas compressors: a low pressure compressor and a high pressure compressor.

In this example which is illustrated in Fig. 3, unprocessed oil (which consists of oil, gas and water) enters the stage 1 separator. The gas that is separated from the stage 1 separator goes to the low pressure gas compressor. The leftover oil flows to the stage 2 separator, where the remaining gas is separated from the oil. This gas is also sent to the low pressure gas compressor. The water that is separated during the process is sent to the water treatment equipment. The refined oil is stored and/or exported after leaving stage 2 separation. After leaving the low pressure gas compressor, the gas is sent to the high pressure gas compressor, where it is then exported.

As illustrated in Fig. 3, cluster 1 is responsible for monitoring the compressors, cluster 2 is responsible for the separators, cluster 3 monitors the water treatment equipment and cluster 4 monitors the oil storage tanks.

Alternatively, if there were some sort of hierarchical sub-clusters on a lower level, cluster 1.1 can monitor the low pressure gas compressor and cluster 1.2 can be responsible for the high pressure compressor. Some facilities will have more

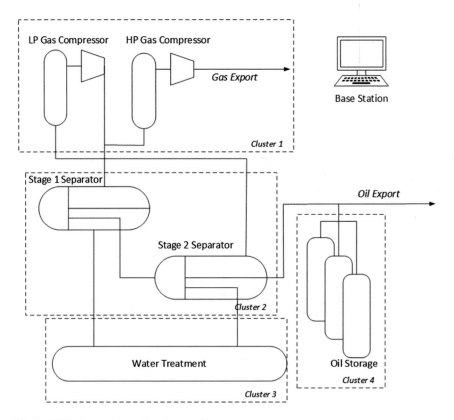

Fig. 3 WSN clusters in an oil and gas refinery

than 1 piece of given equipment. For example, if it is assumed that a regular sized oil storage tank can be cluster 1.1, and that there are three oil storage tanks in a refinery, it can then be said that cluster 1.1.1 monitors the first storage tank, 1.1.2 monitors the second, and 1.1.3 monitors the third. This demonstrates scalability of the wireless sensor network. In this case, it would be simple to add one or more storage tanks, or remove an obsolete or damaged storage tank.

6   Conclusions and Future Work

In this paper, a model was proposed to embed intelligent agents on sink nodes and manage clusters. The agent architecture and deployment was examined and illustrated in an example for an oil and gas refinery. This type of embedded deployment has illustrated several challenges. The most important one can lay in fault recovery; for example, how the multi-agent system reacts when a sink node fails.

There are also challenges present when inter cluster communication is present, due to the embedded agents. In a cloud-based system, a centralized deployment has agents that have access to most of the network to negotiate with the other sink nodes. In an embedded system, a sink node may not be in range of other clusters in order to borrow sensor nodes, amongst many other challenges. A new protocol for network discovery and integration needs to be developed for the plug and play style of introduction of new sensor nodes. As previously mentioned this is a key challenge for industrial systems and can be linked to it being an under-researched field.

Currently, an embedded agent-managed cluster environment is being developed. This environment has the agents implemented on higher-powered micro-computers (such as the Raspberry Pi), and uses the ZigBee RF network protocols using xBee transmitters. The anchor nodes consist of simple controllers, such as PLCs. This hardware will provide simulated data back to a SCADA system and will allow many situations to be examined.

References

1. Broy, M., Schmidt, A.: Challenges in engineering cyber-physical systems. Computer **2**, 70–72 (2014)
2. Lasi, H., et al.: Industry 4.0. Business & information. Syst. Eng. **6**(4), 239–242 (2014)
3. Callaway, E.H.: Wireless Sensor Networks: Architectures and Protocols. CRC Press (2003)
4. Wooldridge, M.: An Introduction to Multiagent Systems. John Wiley & Sons (2009)
5. Leitão, P., Karnouskos, S. (eds.): Industrial Agents: Emerging Applications of Software Agents in Industry. Morgan Kaufmann (2015)
6. Hla, K.H.S., Choi, Y.S., Park, J.S.: The Multi Agent System Solutions for Wireless Sensor Network Applications. Agent and Multi-Agent Systems. Technologies and Applications, pp. 454–463. Springer, Berlin, Heidelberg (2008)
7. Karlsson, B. et al. (2005). Intelligent sensor networks, an agent-oriented approach. In: Workshop on Real-World Wireless Sensor Networks

8. Tynan, R., Ruzzelli, A.G., O'Hare, G.M.P.: A methodology for the development of multi-agent systems on wireless sensor networks. In: 17th International Conference on Software Engineering and Knowledge Engineering (SEKE'05). Taipei, Taiwan, Rep. of China, 14–16 July, 2005
9. Gholami, M., Taboun, M., Brennan, R.W.: Comparing alternative cluster management approaches for mobile node tracking in a factory wireless sensor network. In: 2014 IEEE International Conference on Systems, Man and Cybernetics (SMC). IEEE (2014)
10. Ningxu, C., et al.: Application-oriented intelligent middleware for distributed sensing and control. Syst. Man Cybern. Part C Appl. Rev. IEEE Trans. **42**(6), 947–956 (2012)
11. Savazzi, S., Guardiano, S., Spagnolini, U.: Wireless sensor network modeling and deployment challenges in oil and gas refinery plants. Int. J. Distrib. Sensor Netw. (2013)
12. Gil, P., Santos, A., Cardoso, A.: Dealing with outliers in wireless sensor networks: an oil refinery application. Control Syst. Technol. IEEE Trans. **22**(4), 1589–1596 (2014)
13. Hevard, D.: Oil and gas production handbook (2006)

Author Index

© Springer International Publishing Switzerland 2016   339
T. Borangiu et al. (eds.), *Service Orientation in Holonic and Multi-Agent Manufacturing*, Studies in Computational Intelligence 640,
DOI 10.1007/978-3-319-30337-6

Printed in the United States
By Bookmasters